The Compact Kernel
of a Metabolic Flux
Balance Solution Space

Concepts, Algorithms and
Implementation

The Compact Kernel
of a Metabolic Flux
Balance Solution Space

Concepts, Algorithms and Implementation

Wynand S Verwoerd

Lincoln University, New Zealand

W⟲ World Scientific

NEW JERSEY · LONDON · SINGAPORE · BEIJING · SHANGHAI · HONG KONG · TAIPEI · CHENNAI · TOKYO

Published by

World Scientific Publishing Co. Pte. Ltd.

5 Toh Tuck Link, Singapore 596224

USA office: 27 Warren Street, Suite 401-402, Hackensack, NJ 07601

UK office: 57 Shelton Street, Covent Garden, London WC2H 9HE

British Library Cataloguing-in-Publication Data
A catalogue record for this book is available from the British Library.

**THE COMPACT KERNEL OF A METABOLIC FLUX BALANCE
SOLUTION SPACE**
Concepts, Algorithms and Implementation

ISBN 978-981-125-583-0 (hardcover)
ISBN 978-981-125-584-7 (ebook for institutions)
ISBN 978-981-125-585-4 (ebook for individuals)

For any available supplementary material, please visit
https://www.worldscientific.com/worldscibooks/10.1142/12821#t=suppl

Typeset by Stallion Press
Email: enquiries@stallionpress.com

Dedication

To Nona, for her lifelong loving support of my research career, without which this book could not have been written. I hope it will be a tangible expression of my appreciation for the disruptions to her own life and pursuits as she accompanied me on many journeys across the globe and through life.

Contents

List of Tables

List of Figures

Chapter 1

Introduction

1.1 Constraint-Based Analysis of Steady States

Constraint-based analysis [1–4] has become a standard tool in modelling the metabolism of living cells.

A fundamental point of departure is that there is a network of biochemical reactions, each catalysed by specific enzymes, that receive nutrients from the cellular environment and chemically transform these into metabolic products needed by the cell and/or waste products returned to the environment. Intermediate products are passed on from one reaction to another, and these connections embody the layout or structure of the network. The cellular genome selects which enzymes can be produced and thus broadly determines the subset of biochemical reactions that make up the network.

The regulatory network in a cell exerts more detailed control of the metabolism, by dynamically determining the quantity of enzymes available in response to factors such as external stimuli, the particular cell type in a multicellular organism, the current stage of the cell cycle and so on. As a result, the chemical flux through any particular reaction in the network can change. So, the metabolic state of a cell is specified by stating the set of flux values associated with every reaction in the network at a particular moment.

Mathematically, this can be considered to be a point in an abstract flux space, which is described by a set of mutually orthogonal coordinate axes, each of which represents the flux through a particular biochemical reaction in the network. This point is mathematically specified by its position vector, usually referred to as the flux vector. Metabolic changes can be visualised as a trajectory in flux space, which is traversed by the flux vector giving the momentary state of the cell during its life history. The actual trajectory can be expected to be influenced not only by the abovementioned factors, but since each cell is slightly different, each cell in a colony or tissue can be expected to traverse a unique trajectory.

The main aim of the constraint-based approach is to model the limits imposed on such trajectories by physical and chemical principles, rather than the details of where the metabolism of a particular cell is located at any moment and on what trajectory. The main constraints on metabolic states stem from the stoichiometry of each reaction, which determines ratios of different metabolite flows; thermodynamics, which determines the direction of each chemical reaction, and the metabolite concentrations and activity level of each enzyme that limits maximal fluxes that can be realised. As a result, metabolic trajectories are constrained to stay within a *feasible* region of the flux space, and this is referred to as the 'Solution Space' (SS). 'Feasible' in this context means that the region satisfies all the physico-chemical constraints.

The equations of motion for the flux point are a set of time-dependent first-order differential equations that express conservation of mass in each chemical reaction. But to avoid that level of detail, constraint-based analysis usually makes the assumption of a steady metabolic state. When all time derivatives are accordingly set to zero, the equations reduce to the matrix equation

$$S \cdot f = 0 \qquad (1.1)$$

Here, S is a matrix containing the stoichiometric coefficients of each reaction in each column, corresponding to the metabolite in each row. The vector f is the flux vector mentioned above. Each row of S constitutes a stoichiometry constraint on feasible values of f.

A thermodynamic constraint that fixes the direction of reaction i to the forward direction gives the inequality $f_i \geq 0$ and a maximal flux value is similarly expressed as $f_i \leq \phi$. These are referred to as range constraints in what follows.

The steady state assumption can be justified by noting that chemical equilibration is much faster than biological cellular processes, so that kinetic aspects of metabolism can plausibly be considered to be a sequence of transitions between steady states. In this sense, the metabolic trajectory is restricted to the same SS by physico-chemical constraints as any steady state so that Eq. (1.1) remains appropriate to characterise the feasibility region for trajectories.

An important constraint-based approach is Flux Balance Analysis (FBA) [3], which makes use of optimisation methods. Here, the SS is further constrained by optimising an objective. For example, for microbes, the biomass growth rate is often maximised. More elaborate versions may combine several objectives. Generally, this may be considered as a way to encapsulate the overall effects of the cellular regulation into a single maximised value, or a small set of such values.

Even when this objective value is kept fixed to its optimised value, the feasible region can be expected to remain extensive, and can still be regarded as an SS rather than just a fixed single flux point. This reflects the biologically plausible situation that, for example, different microbes in a colony can attain the same biomass growth rate even when using slightly different nutrient mixes and/or metabolic routes. This fact is somewhat masked by the fact that the computational methods, such as Linear Programming (LP) [4] that is typically used to perform FBA, delivers only a single flux point. It is nevertheless generally understood that this is merely a representative solution, and various ways further discussed below have been employed to more fully describe the SS.

The subspace that contains all feasible solutions, that in addition produce the optimised objective values, might more properly be referred to as the Objective Space (OS). The OS generally has fewer dimensions than the SS. However, for practical purposes, the distinction between SS and OS is usually unimportant as they both represent multidimensional polytopes and mostly allow the same types of characterisation and analysis. The term SS is mostly used below in a generic sense to refer to either of these concepts.

1.2 Characterising Solution Spaces

The number of reactions and metabolites that make up the metabolic network can vary from several hundred for simple microbes to several thousand for more complex organisms. This means that the associated flux spaces have correspondingly large dimensions.

As formulated in Eq. (1.1) and its discussion, the physico-chemical constraints are all mathematically represented by equations and inequalities that are linear in terms of fluxes. So, the SS they describe is bounded by flat hyperplanes and is thus a polytope, in a many-dimensional space. Usually, the FBA objectives that are used are also linear, so even the OS remains a polytope, perhaps in a space with somewhat fewer dimensions.

There are two equivalent ways to specify a polytope as a geometrical object. Most intuitive is to specify the positions of all vertices. In two dimensions, polygons like triangles, rectangles, hexagons and so on are easily described that way, as are polyhedra like cuboids, octahedra and so on in three dimensions. This is sometimes referred to as the *V*-representation of a polytope.

The alternative is to specify only the sides explicitly while vertices are at the implied intersections of these. For example, the vector equation

$$c \cdot f \le v \qquad (1.2)$$

defines the half-space of all points *f* that satisfy the inequality, given a vector *c* and a scalar value *v*. The straight line, flat plane or hyperplane that borders the half-space, is oriented orthogonal to *c*. Restricting *c* to be a unit vector, the value *v* gives the perpendicular distance from the origin to the hyperplane. The interior of a polytope is the intersection of all the half-spaces defined by all its sides, that is, the collection of points that simultaneously satisfy a set of inequalities of the form of Eq. (1.2), one for each side. This set of inequalities has the matrix form

$$C \cdot f \le V \qquad (1.3)$$

where C is a matrix and V a vector of values.

If the coordinate origin is chosen inside the polytope, the distance to each side is a positive number, or zero if the origin falls on the corresponding border. An example of the half-space representation in 2D is shown in Figure 1.1.

Similarly, the interior of a cube can, for example, be defined as the set of 3D points that simultaneously satisfies a set of six such inequalities.

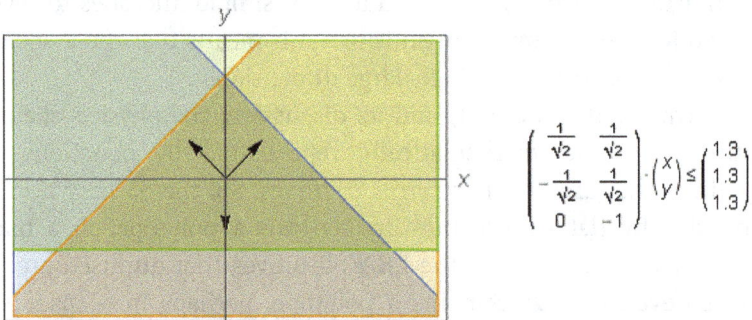

$$\begin{pmatrix} \frac{1}{\sqrt{2}} & \frac{1}{\sqrt{2}} \\ -\frac{1}{\sqrt{2}} & \frac{1}{\sqrt{2}} \\ 0 & -1 \end{pmatrix} \cdot \begin{pmatrix} x \\ y \end{pmatrix} \le \begin{pmatrix} 1.3 \\ 1.3 \\ 1.3 \end{pmatrix}$$

Figure 1.1 H-representation of an equilateral triangle. Each arrow shows a unit constraint vector, perpendicular to a corresponding constraint line at a distance given by the constraint value. A different shading colour for each line indicates the half plane of points that satisfy the constraint. The triangle where all of these overlap defines the feasible polygon defined by the matrix equation shown.

This half-space specification is sometimes called the H-representation of a polytope.

To cast this geometric interpretation of Eq. (1.2) into the linear optimisation context used for constraint-based analysis, it can be considered to be a constraint equation that limits the vector f to the half-space defined by a *constraint vector c* and a *constraint value v*. Correspondingly, Eq. (1.3) involves the *constraint matrix C* and *values vector V*.

One advantage of the H-representation is that it is straightforward to use it for a polytope that is unbounded in some directions, such as an infinite length triangular prism, open at both ends, which does not have any vertices at all! Another advantage is that the number of vertices may grow exponentially with the number of dimensions, even for shapes where the number of sides only grow linearly.

In N dimensions, the smallest number of sides that can give a closed volume is that of a simplex, namely $(N + 1)$ sides. In low dimensions, this is intuitively clear; for example, the 2D simplex is a triangle and the 3D simplex is a tetrahedron. Each linear constraint fixes one degree of freedom. Since a vertex is a geometrical point, all its N degrees of freedom are fixed, and so it is defined by the intersection of N out of the $(N + 1)$ sides of the simplex. Therefore, a simplex also has $(N + 1)$ vertices in any number of dimensions.

But consider a hypercube with sides perpendicular to the coordinate axes. That clearly has $(2N)$ sides, and 2^N vertices. For 2D and 3D, the number of sides and vertices are similar, but in high dimensions, the proliferation of vertices makes it much easier to deal with a side-oriented specification than with a vertex-oriented description.

To see how these remarks apply to the metabolic SS, it is instructive to compare linear optimisation with linear algebra. In the latter domain, a standard problem is to find all solutions to a set of linear equations, given by

$$A \cdot x = b \tag{1.4}$$

Whenever the coefficient matrix A of the set of equations is rank-deficient, it has an SS that is a subspace of the multidimensional space spanned by the axes associated with all variables contained in vector x.

A standard solution method [5] for such problems makes use of the Moore–Penrose pseudoinverse matrix A^+. If there are solutions, they are given by

$$x = A^+ \cdot b + (I - A^+ \cdot A)w \qquad (1.5)$$

Here, w is an arbitrary vector and I is the identity matrix. The properties of the pseudoinverse are such that matrix $(I - A^+ \cdot A)$ is a projection matrix on the null space of A. Hence, the second term of Eq. (1.5) is an arbitrary vector in the null space of A, which has fewer dimensions than the column space of A, to which x belongs.

So, Eq. (1.5) implies that the SS of this linear equation problem is fully specified by giving a particular solution (vector $A^+ \cdot b$), and the basis vectors of the null space of A.

The linear optimisation problem is an extension of solving linear equations, where some or all equations become inequalities and there is in addition an objective that is optimised. LP as employed by FBA delivers the particular solution, but not the basis vectors.

The latter component is the task addressed by a second major class of tools used in constraint-based analysis, namely Extreme Pathway Analysis [6] (ExPa) or the closely related [7] Elementary Mode Analysis [8] (EM). Observing that given two feasible flux combinations, any flux that is intermediate between those will also be feasible, it follows that the SS polytope is in fact a *convex* polytope. The finite basis theorem [4] shows that any point in a convex polytope is specified as a convex combination of vectors that belong to a finite convex basis, consisting of the position vectors of the polytope vertices. A convex combination means that all combination coefficients are in the range 0 to 1 and they all add up to a value 1.

In the metabolic modelling context, each SS polytope vertex defines an extreme pathway, and any feasible flux point is a convex combination of these extreme fluxes.

From this mathematical perspective, knowledge of all the extreme pathways fully determines the SS. If the optimised flux determined by FBA is added as an additional linear constraint, the OS can also be fully specified in the same way.

There is, however, a snag. Even for a modest network representing just the central metabolism of *Escherichia coli*, more than half a million elementary modes have been detected [9]. From a geometric perspective, this is just a manifestation of the exponential dependence of the vertex count on the SS dimension as mentioned previously. A network perspective provides a more detailed explanation, that there is a 'combinatorial explosion' because of the multiple ways that metabolic pathways through one part of a network can be connected to those through another. An analysis that estimates the number of ExPas based on network parameters [10] yields estimated counts of the order of 10^{18} for an *E. coli* model with 904 fluxes, and 10^{29} for a *Homo sapiens* model with 3311 fluxes.

This explosion of elementary modes count (i.e., vertices) with increasing network size, not only makes it computationally prohibitive to find all modes, but also means that extracting useful information and interpretation from them can become intractable for genome-scale metabolic networks (GSMs).

The combinatorial concept suggests that the problem can be dealt with by splitting the large network into smaller subnetworks that can be separately studied. For example, structural features such as node connectivity can be combined with biochemical insights [11–12] to identify modules; and many other methods have been proposed [13] as well. However, a main aim of modelling genome-scale networks is to provide a holistic, systemic description, in contrast to the conventional biochemical approach of identifying individual pathways that correspond to biological processes in isolation. Inevitably, modularisation sacrifices the holistic view, even if to a lesser degree than pathways do.

A related idea is to reduce the network size by discarding reactions and metabolites not considered crucial so the network is reduced to an essential core. For example, a so-called *core network* for the *E. coli* microbe [14] has been compiled and is available from the BiGG [15] online depository, consisting of only 263 constraints on 95 flux variables rather than the 7302 constraints on 2712 fluxes of a more realistic full GSM [16] for *E. coli*. That core model is intended mainly for educational purposes, but a subsequent model designated as EColiCore2 [17] has been extracted from the full GSM by a purpose-made network reduction algorithm [18] called NetworkReducer, and is intended to allow the application of computationally intensive metabolic analysis that is not possible for the GSM. This was further elaborated by a mixed-integer LP method designed to guarantee the minimality of the network reduction [19]. Graph-based reduction methods have also been proposed [20]. Reduced network models for human metabolism have been produced to model particular cell types such as red or white blood cells, or for example, the mitochondria-centred core metabolism [21].

A somewhat different approach to network reduction has been proposed [22] under the name NetRed. It is based on a matrix transformation of the stoichiometry matrix, which maintains reactions that were specifically user nominated while others are condensed into equivalent reactions in which intermediate metabolites are suppressed.

The major focus of the work presented below, is to extract the **Solution Space Kernel** (SSK) of a metabolic network. The term 'kernel' as introduced in Section 1.4 is chosen to avoid confusion with *core networks* mentioned earlier. The work here is not about network reduction. Instead, it is aimed at characterising a specially defined, compact subspace of the SS of any metabolic network, whether it be a full GSM or a reduced version of that. For example, in a later chapter, the SSK of the various *E. coli* core networks will be discussed in conjunction with the SSK of the genome-scale *E. coli* network.

In much of the contemporary work, the focus is on extracting information that is useful for bioengineering from the elementary modes, such as identifying minimal cut sets (MCS) [23–25]. This can allow determination of only a subset of the EMs [26, 27] and so the emphasis on fully characterising the SS is reduced.

However, there is one commonly used method that gives an easily accessible estimation of the SS – flux variability analysis (FVA) [28]. This is both simple to implement and intuitive to understand, although quite demanding computationally. Briefly, it adds one linear constraint to Eq. (1.1) that keeps the objective value fixed at the optimal value calculated by FBA, and then does a pair of separate LP calculations for each flux vector component, to establish the minimum and maximum feasible values of each reaction flux.

1.3 FVA – Strengths and Weaknesses

FVA results in a list of limits within which each flux can vary, which is easy to grasp. Even geometrically, it is straightforward to visualise the result as a cuboid region in N-dimensional space, within which the SS is fully contained, that is, a bounding box for the SS.

A particularly striking aspect of the limits typically obtained is that a large number, often around 50%, of fluxes in fact remain fixed in value throughout the SS.

Flux values can for physical reasons obviously not increase without limit. Although actual limits reflecting the activity of enzymes, the finite amounts of metabolite molecules in the cellular volume and so on are usually unknown, most available metabolic models contain artificial upper flux limit values that are arbitrarily chosen to be large enough not to restrict any realistic flux values. This is a device presumably to accommodate LP solver implementations that require finite limits, although the underlying LP algorithms do not require such limits.

A second strikingly large group of flux limits obtained in a typical FVA calculation are fluxes for which one or both limits are equal (or comparable) to this artificial value. Such fluxes can be considered to be unbounded by the model constraints, and only attain finite extremes because of the artificially imposed limits.

That leaves a usually small group of fluxes for which the range of variation is non-zero but within physically plausible limits. These 'variable' fluxes arguably are the ones that reflect the meaningful links between flux values, as embodied in the metabolic model.

Identifying these three groups of fluxes (fixed, variable and effectively unbounded) is a major benefit of FVA, and provides a starting point for further refinement of the SS concept as explored below. But first, some theoretical and practical limitations of the technique also have to be recognised.

A severe theoretical limitation is that, perhaps counter-intuitively, the FVA bounding box actually gives virtually no information about a high-dimensional SS apart from a broad indication of its locality and overall extent in the flux space. In 2D and 3D examples, the situation is simpler: the bounding box of a polygon or polyhedron is a reasonable first approximation to visualise its shape and orientation, since the polytope occupies a sizeable fraction of the bounding box volume. But this relationship breaks down in high dimensions.

To appreciate that, consider a hypersphere with radius R centred at the origin, to approximate a highly symmetric convex polytope. This polytope with a very large number of faces (and vertices), is constructed in such a way that among the many vertices there is a pair of symmetrically opposite vertices located on the positive and negative sides of each coordinate axis, all at a distance R from the origin.

The bounding box in this case is a hypercube with opposing sides separated by lengths $2R$, or a hypervolume $V_C = (2R)^N$. The hypersphere with radius R inscribed in the bounding box encloses the described polytope by construction, so it has a hypervolume larger than that of the polytope.

The sphere hypervolume V_S is given by [29]

$$V_S = \frac{\pi^{N/2}}{\Gamma\left(1+\dfrac{N}{2}\right)} R^N \approx \frac{1}{\sqrt{N\pi}} \left(\frac{2\pi e}{N}\right)^{N/2} R^N \qquad (1.6)$$

where the last part is the large N approximation derived from the Stirling formula, and gives an upper bound to V_S. It follows that

$$V_S/V_C \xrightarrow[\infty]{N} 0 \qquad (1.7)$$

In other words, the inscribed hypersphere and therefore also the polytope it encloses, occupies a negligible fraction of the volume of the bounding box for large N values. This statement applies even more to a realistic SS polytope that is less symmetric and has fewer vertices, as that will occupy a smaller fraction of the volume of the hypersphere inscribed in its bounding box.

Applying this reasoning to even a very small metabolic network with only 10 reactions, Eq. (1.6) shows that if the fraction of the FVA bounding box that is occupied by the SS polytope is estimated to be in the ball park of V_S/V_C, this value is $<10^{-6}$ for a value as small as $N = 10$. It follows that the FVA bounding box only gives a very weak characterisation of the actual SS in this example, and for larger networks its relevance becomes exponentially worse.

Turning to practical problems, the most obvious problem with FVA is that it can become very slow to calculate for medium to large networks, since it requires $2N$ individual LP calculations, and each of these scales proportional to a power of N typically between two and three, depending on the LP algorithm chosen. This problem has been addressed by the development of a FastFVA algorithm [30]. In that approach, the outcome of each LP calculation is used to improve the starting point of the next, sometimes called a 'hot start' approach. In the quoted study trials for networks of different sizes and two different LP solvers were reported, and computing time reductions by a large factor that ranges between roughly 20 and 200 were found.

There are, however, also more insidious numerical problems. These are best appreciated by considering a case study, using a metabolic model [31], updated in Sun et al. (2009) [32] for *Geobacter sulfurreducens*. This is a medium-sized GSM with 705 metabolite constraints and a total of 940 flux variables after replacing reversible reactions with an equivalent pair of oppositely directed reactions. In the subsequent work, it turns out that the Geobacter SS shows a variety of features typical of most models, whereas smaller models often only show a subset of such features. It is used as a demonstration model throughout this work.

The calculations presented here were done using two different LP solvers that are integrated in the *Mathematica* software package [33], namely one using the Simplex algorithm and the second using an Interior Point (I.P.) algorithm. Both of these solvers are capable of handling both finite and infinite bounds on variables, so were applied to the original model that contains artificial upper bounds of flux value 1000, as well as the same model with all artificial bounds removed. To further investigate the effect of artificial bounds, another version where the bounds were set to a value 10,000 was also calculated with both solvers, giving a total of six sets of calculations.

These trials are chosen to test two premises commonly assumed in FBA and FVA modelling, the second implicit in the very formulation of models with artificial flux bounds:

- LP solver algorithm choices may influence convergence, scaling with problem size and hence computation times, but are equivalent regarding the *value* of the optimised objective (even though different optimised fluxes may result where the constraint problem is underdetermined).
- The actual value of the flux upper bound is immaterial as long as it is well above the range of physically tenable flux values.

For the FBA calculation that maximises the biomass growth in the chosen Geobacter model, these assumptions are indeed borne out. All three Simplex calculations, carried out with an LP convergence tolerance of LPtol $= 10^{-6}$, agree to 16 significant figures on an objective value of 0.659544 h^{-1}, taking 3.5 sec of computing time for the bounded cases and 1.1. sec for the unbounded one. The I.P. method only needs <0.08 sec in all three cases, but it is an iterative method and can only converge within a tolerance LPtol $= 5 \times 10^{-5}$. Even so its objective values are respectively 0.659547, 0.659545 and 0.659545 h^{-1}, deviating from the Simplex values by $<3 \times 10^{-6}$.

Turning to FVA, it requires 1880 separate LP calculations in each of the six cases, and regarding the assumptions, the picture is quite different from FBA. Starting with the I.P. calculation of flux ranges for the original bounded model, the first sign of trouble is that many of the individual LP calculations fail to converge unless the iteration tolerance is increased by more than two orders of magnitude to LPtol = 0.003. Even at this level, one case fails completely to produce a solution. Simplex performs better, achieving solutions for all cases. But even so, a more serious problem is observed for both LP methods: in more than 10% of cases, the calculated lower bound is higher than the calculated higher bound. Bearing in mind that the lower and upper bound calculations for the same flux, is exactly the same LP problem except for changing the sign of the objective vector, this logical contradiction can be ascribed to numerical inaccuracies in each of such a calculation pair. If so, the maximum over the 120 cases where this discrepancy occurs can be used as an estimate of the flux accuracy. For I.P., this works out at 0.006 and even worse at 0.014 for Simplex. These values are roughly four orders of magnitude worse than what was achieved for FBA.

Repeating FVA for the remaining four trial cases, the number of inconsistent values is similar in each case, but the extent of the discrepancies is mostly worse, except for the Simplex, unbounded trial that yields a better maximal discrepancy value at 0.0006.

A main feature of the FVA analysis is to distinguish between fixed fluxes, variable fluxes and unbounded fluxes. Limited accuracy also sets a limit on the range of calculated flux variation that can be considered to be negligible, and thus to denote a flux as fixed. In order to use a uniform criterion over the six trials, the worst-case accuracy was taken to establish the numerical fixed value tolerance as *fixtol* = 0.025, that is, any larger range is considered to indicate a variable flux. The resulting number of fluxes allocated to various categories by the six trials, are shown in Table 1.1.

Table 1.1: Summary of flux classification counts from Simplex and I.P.-based FVA calculations for the *Geobacter* example model, with/without artificial flux bounds.

	Trial	Comp Time	FIXED	VARIAB (lower)	VARIAB (upper)	UN BOUND
SIMPLEX	Art. Bnd 1000	6084	438	40	67	395
	Art. Bnd 10,000	6821	433	45	63	399
	Unbounded	2100	427	51	0	462
INTERIOR POINT	Art. Bnd 1000	110	439	40	77	384
	Art. Bnd 10,000	256	438	41	63	398
	Unbounded	117	439	40	0	461

While all the trials present the same broad picture that there are somewhat more than 40% of all fluxes in each of the fixed and unbounded categories and 5% to 10% are variable, there are significant discrepancies in the detailed counts shown in the table. It might be supposed that these arise from the limited accuracy, but inspection of individual cases reveals that only a few cases can be accounted for in this way. The vast majority are in fact direct contradictions.

For example, among the fixed fluxes, only 426 are identified as such by all six trials. There are six fluxes that are fixed at the same value by five of the trials, but are found variable by the unbounded Simplex calculation. Of these, three are because of a different lower bound and three because of a different upper bound, and these differences are of order unity, which is far beyond any ambiguity arising from accuracy limitations and so are downright contradictions. Another set of five fluxes are similarly found variable by both the large bound and unbounded Simplex trials while fixed according to the other four trials. An even more extreme contradiction is found in one further case, where a flux is fixed according to all three I.P. calculations, and unbounded according to all the Simplex trials. The I.P. fixed flux classifications are mostly mutually consistent, but even here, there is one flux that is found variable (albeit within a small range) with the large artificial bounds, while fixed in the other cases.

Turning to the variable fluxes, Table 1.1 presents them as two separate groups, a lower and higher group. The reason is seen in Figure 1.2, which shows the distribution of FVA upper limits for the flux ranges of variable fluxes for the two choices of artificial bounds that were trialled, as calculated by I.P. LPs.

The upper group of variable flux limits are distinguished by the fact that they shifted upwards from values peaking around 100 to new values peaked around 1000, as the artificial bound was similarly increased by a factor of 10. The lower group, in the range from 0 to 10, by contrast remained essentially unchanged by the adjustment of the artificial bound.

This lower group, incidentally, also represents values that are physically plausible, for example, by comparison to the observed [34] maximal nutrient (acetate) uptake rate of 18 mmoL/gDW/h for *Geobacter sulfurreducens*.

The fact that the quite large upper group of more than 60 fluxes shift in accordance with the artificial bound value, disproves the validity of

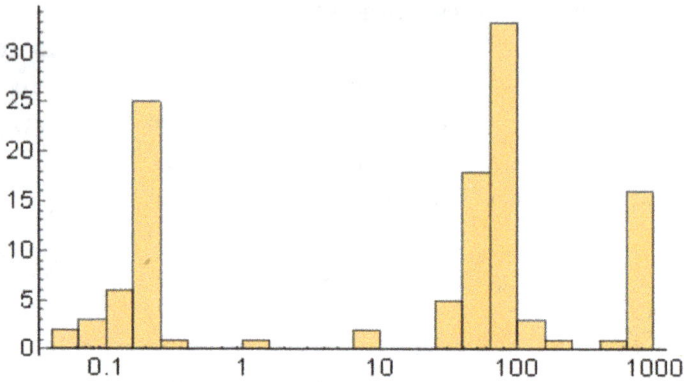

(a)

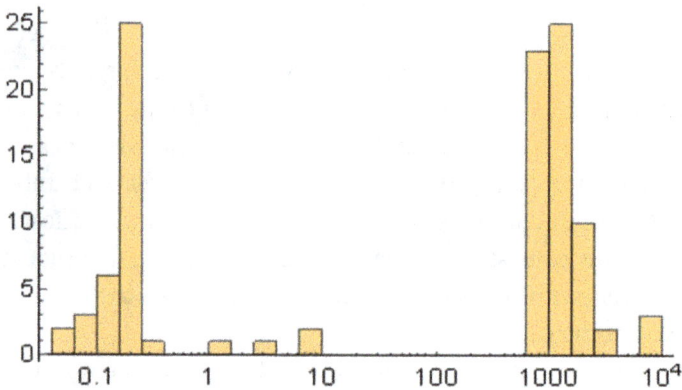

(b)

Figure 1.2 Histogram of FVA variable flux upper limits for (a) artificial upper flux bound = 1000 and (b) artificial upper flux bound = 10,000 as calculated by Interior Point LP. Flux values on the horizontal axes are in mmol/gDW/h and the vertical axis shows flux counts.

the second assumption in the previous bullet point list. Its applicability might be rescued if it is possible to allocate all these fluxes to the group considered to be in principle unbounded, which is in accordance with the fact that as seen in Table 1.1, the higher group vanishes completely when the artificial bounds removed from the model. But as seen from Figure 1.2(a), it is not straightforward to choose a dividing line between the lower and higher groups for the artificial bound value = 1000 used for many models. Moreover, the table also reflects that the similar Simplex calculations display the same general behaviour but with different allocations of variable flux group membership.

The simplest solution to this dilemma is to discard the second assumption and perform all FVA calculations without any artificial bounds. But that still leaves the inconsistencies and contradictions between different LP solvers, so the first assumption is also problematic, and not so easy to dispose of.

Problems with accuracy and convergence in LP-based metabolic models have been pointed out by previous authors and remedies proposed have included improving numerical routines, for example, by using quadruple precision LP solvers [35] or even exact arithmetic solvers [36], although others have questioned the need for these [37–38].

The case study above and the contrast between FBA, where no problems occurred, and FVA, which suffered from convergence problems, inaccuracies, inconsistencies and downright contradictions between supposedly equivalent trial calculations, suggests to us that the problem may be deeper than just numerical accuracy.

Instead, the FVA appears to be an inherently ill-posed problem in a way that FBA normally is not. Why this should be so, may look mysterious, as the two problems are very similar. They are optimisation problems with the same flux variables subject mostly to the same constraints, except that the FBA objective becomes an additional constraint for FVA, and then each flux variable in turn becomes a new objective to be either minimised or maximised.

A clue for why this apparently minor change might upset the well-posed nature of the FBA problem comes from the work to be presented in subsequent sections of this book. In broad outline, that involves a sequence of vector space projections on lower dimensional subspaces. In particular, in this approach, the fixed FBA objective value is enforced first and that defines such a subspace. The range constraints that applies to the flux variables, are then subsequently projected down to this subspace. After the projection, constraint vectors (in the sense of the term defined in connection with Eq. (1.2)) that were originally orthogonal, may become parallel or anti-parallel, or project to zero. Any slight numerical error in such a projection that misaligns two constraint hyperplanes that are in principle parallel, will cause them to intersect and hence introduce a finite limit to a direction in flux space that should in principle be unbounded. Conversely, one way that a fixed flux can arise is where a sufficient number of constraint hyperplanes intersect at the origin (e.g., three lines in two dimensions). If one of these infinitesimal constraint vectors acquires the wrong arithmetic sign because of inaccurate projection, it can fail to be effective and allow a flux variable that is in principle fixed, to become unbounded. These types of instabilities are exactly the inconsistencies between the various trials that arose in the case study.

Such instabilities are hidden from view in the standard LP formulation of FVA, because no vector projections are done explicitly. In fact, the flux range constraints are not dealt with (e.g., in the Simplex LP) as explicit constraints on the same basis as other constraints, but are instead incorporated in the way that the iterations towards the optimal solution are performed. We surmise that this projection is nevertheless implicitly involved in enforcing the fixed objective value constraint that is added by FVA, and that this causes the observed problems with FVA.

Clearly increased precision may ameliorate the problems, but a better approach is to find a different description of the SS that avoids them altogether. This view is reinforced by the fact that even if executed perfectly, the FVA bounding box would in any case give a poor characterisation of the SS because of the discrepancy in their hypervolumes. Developing such a new descriptive framework is the purpose of this study.

A recent review of network reduction methods [13] reinforces the view that the new framework presented here is completely different from what is currently available in the literature.

1.4 A New Description of FBA Solution Spaces

Regarding the SS, the main takeaway from FVA is neither its calculated bounding box that vastly overestimates the extent of the SS, nor the detailed range that it allocates to each flux and which proves to be somewhat unreliable, but the principle of classifying fluxes into three groups: fixed, variable and unbounded.

Of these, the variable fluxes are of most interest because they represent the opportunities to bioengineer the metabolism of a cell for a particular purpose, without disturbing a primary objective such as optimised growth.

Fixed fluxes represent a counterpart to this, namely the non-negotiable part of the flux space.

Unbounded fluxes are more problematic, and this is not helped by assigning them artificial bounds. Clearly, there are always physical limits to flux values, so the absence of explicit flux bounds generally indicates incomplete knowledge. This lack of knowledge may not matter in some circumstances. If limits are known on some fluxes, such as maximal nutrient uptake measurements, for example, the range of variation of other fluxes may be constrained by the way that fluxes are linked by the metabolic network and its stoichiometry constraints. In particular, a known fixed objective such as the biomass growth rate could in principle constrain all other fluxes to finite ranges.

The following almost trivial example illustrates from a geometric perspective how this can happen. Consider the case of just three fluxes f_1, f_2 and f_3 that are restricted to positive values by firm knowledge of the direction of their associated chemical reactions, but where there are no known bounds on the values of the fluxes. This would be represented by the coordinate axes of the positive octant of a 3D flux space. Suppose that the objective is the flux combination $(f_1 + f_2 + f_3)$ and its optimised value is F. This defines a plane in the flux space perpendicular to the constraint vector $(1, 1, 1)$ and located at a perpendicular distance F from the origin. This plane intersects all three axes at the same finite value $\sqrt{3}F$ and so all three fluxes become constrained to the range $(0, \sqrt{3}F)$. Similarly, a fixed value for an objective $(f_1 + f_2)$ would constrain f_1 and f_2 each to the range $(0, \sqrt{2}F)$ while leaving f_3 unbounded.

It follows that the unbounded fluxes cannot be determined merely by inspecting the flux ranges explicitly included in a metabolic model. The FVA analysis goes some way to pinning these down, but even if ignoring its numeric inconsistencies, may only give incomplete information. For example, if FVA finds that f_i and f_j are unbounded, it is possible that only $(f_i + f_j)$ is unbounded while the orthogonal combination $(f_i - f_j)$ has a fixed value or finite range.

Against this background, there are three major tasks:

- An efficient method is needed to calculate fixed fluxes.
- All directions in which the SS are unbounded need to be determined.
- Only then can the main task be attempted, to determine the location, size and shape of the subspace of variable fluxes.

Once the fixed fluxes and their values are known, they can be eliminated from consideration by a vector projection. This simplifies the remaining tasks because they can then be performed in a lower dimensional subspace. It turns out that further reduction steps progressively reduce the dimensions needed to be considered, and can similarly be represented as vector space projections. The example of the fixed flux removal can be used to demonstrate features of all these projections and how they can be mathematically represented in a compact data structure.

Consider a simple 3D SS as shown in Figure 1.3 and defined by six constraint equations, as follows:

$$
\begin{bmatrix}
.707 & -.707 & 0 \\
-.707 & -.707 & 0 \\
.707 & .707 & 0 \\
-.707 & .707 & 0 \\
0 & 0 & -1 \\
0 & 0 & 1
\end{bmatrix}
\cdot
\begin{bmatrix}
f_1 \\
f_2 \\
f_3
\end{bmatrix}
\leq
\begin{bmatrix}
0 \\
0 \\
2 \\
2 \\
-1.5 \\
1.5
\end{bmatrix}
\tag{1.8}
$$

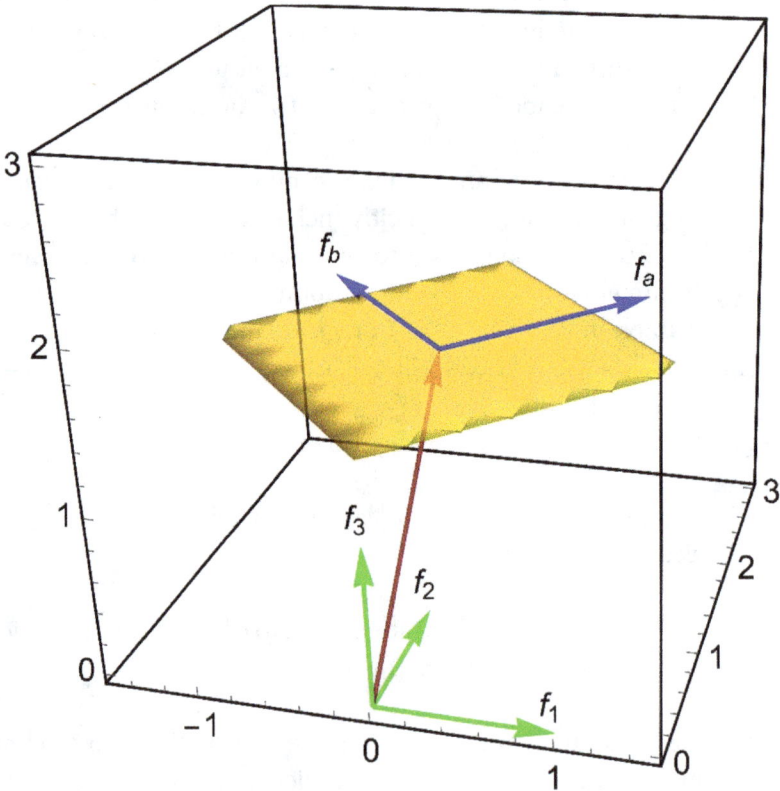

Figure 1.3 Dimension reduction by elimination of a fixed flux from an example solution space in 3D. The SS is the yellow square, the green arrows the three basis vectors of the flux space, and the blue arrows the two basis vectors B of the embedded 2D space that contains the SS. The red arrow is the position O of the 2D origin in the 3D embedding space, chosen as a point interior to the SS.

The last two constraints fix the f_3 flux to a value of 1.5, and the other four define a square region in the plane $f_3 = 1.5$ as shown in Figure 1.3.

Choosing a new origin inside the SS and a set of 2D basis vectors $\hat{f}_a$ and $\hat{f}_b$ relative to this, the SS is equivalently represented by the reduced constraint set:

$$C \cdot f \leq V = \begin{bmatrix} 1 & 0 \\ -1 & 0 \\ 0 & -1 \\ 0 & 1 \end{bmatrix} \cdot \begin{bmatrix} f_a \\ f_b \end{bmatrix} \leq \begin{bmatrix} 1 \\ 1 \\ 1 \\ 1 \end{bmatrix} \tag{1.9}$$

The two equivalent specifications are related by a vector transform T consisting of the position vector O of the new origin, and a matrix B that has the 2D basis vectors (expressed in terms of the 3D basis) as its rows. For the example, and using curly brackets to signify a comma separated list, this is

$$T = \{O, B\} = \left\{ \begin{bmatrix} 0 \\ \sqrt{2} \\ 1.5 \end{bmatrix}, \begin{bmatrix} .707 & .707 & 0 \\ .707 & -.707 & 0 \end{bmatrix} \right\} \tag{1.10}$$

For a given feasible point F, that is, a point inside the SS, the relation between its 2D representation F_{2D} and 3D representation F_{3D} is explicitly given by

$$F_{3D} = O + B^t \cdot F_{2D}$$
$$F_{2D} = B \cdot (F_{3D} - O) \tag{1.11}$$

The first of these equations can be described as an 'upwards' transformation (or point uplift) from the reduced dimension space to the higher dimension, full flux space and similarly the second as a 'downwards' transformation or point downcast.

The transformation proceeds without loss of information in either direction for any feasible point. So, it is possible to analyse the SS and any points inside it in the reduced dimensions without carrying the baggage of fixed flux values around, while fully recovering those values when transforming back to the 3D space. Information about the fixed fluxes is carried in the O vector that forms part of T.

A full specification of the SS, equivalent to the 3D version, is therefore given by listing the set of 2D constraints as a constraint matrix C and values vector V, in a nested list data structure D defined as

$$D = \{\text{constraints, transformation}\} = \{\{C,V\}, \{O,B\}\} \qquad (1.12)$$

With a minor tweak, the example can be used to illustrate how unbounded SSs can be handled in a similar way. Omitting the last constraint, that is, the last row in Eq. (1.8), the SS changes to a semi-infinite square column, stretching to infinity along the vertical axis. The unit vector $\hat{f}_3$ defines the direction in which the SS is unbounded, and is henceforth referred to as a *ray vector*. The cross-sectional shape of the column, in this case a square, is once more defined by the 2D polytope given by Eq. (1.9). This is a simple example of an SSK, a key concept in this study.

To give a general definition, a *Solution Space Kernel* (SSK) is a lower dimensional, bounded subspace of a partially unbounded SS, such that any point P in the SS can be reached from some point P_k in the SSK, by up-transforming P_k as in Eq. (1.11) and adding a positive multiple of an SS ray vector.

Since a ray direction can by definition not point to any vertex of the SS, it follows that the SSK contains all SS vertices and hence all shape information pertaining to the SS.

Both fixed values and ray directions allow specification of an SS by means of a low dimensional, H-represented polytope (as in Eq. (1.9)) and the accompanying projective transformation (as in Eq. (1.10)). But while the actual fixed values are fully contained in the transformation itself, the SSK data structure D that collects these together, needs to be complemented by a list of ray vectors for a complete specification of the SS in the full flux space.

Clearly, the finding of ray vectors is one of the first tasks to be addressed, and it is found that there are several types needing different algorithms for their computation. The ray vector discussed earlier is an example of the simplest type, described as a prismatic ray in the subsequent work. Most, though not all, other ray types can be separated off by means of similar transformations. This gives a progression of reductions in the number of dimensions and constraints that define the SS. Moreover, using matrix algebra, the entire sequence of transformations can be combined into a single one that allows the specification of the final SSK to be cast in the form of Eq. (1.12) and this is one of the final outcomes of this study.

Although the 3D example demonstrates the principle of dimension and constraint reduction, it fails to express the magnitude of the simplification involved when applied to high-dimensional flux models. For example, the *Geobacter* model used as an example in a previous section reduces from 2586 constraints (including both stoichiometry and flux range constraints) on 940 flux variables, to a SSK with 99 constraints on 43 variables. This reduction by orders of magnitude allows a much more detailed analysis of the SS.

A price to pay for this is that the direct connection between fluxes and chemical reactions, or constraints and metabolites, is lost for the variables of the reduced space. But this can be re-established for any flux in the reduced representation, by transforming it upwards using the transformation included in D.

The vector projection that is involved means that although the upwards transformation is lossless for any feasible flux point, the same is not true of infeasible fluxes. Any full flux space point that falls outside the SS hyperplane is projected on it during the down-transformation and remains there during up-transformation, so information about components orthogonal to the SS is lost. For the same reason, the *constraints* in the full flux space cannot in general be recovered by the upwards transformation, unless their constraint vectors were already in the SS hyperplane (as chosen for simplicity in the example). That is, however, of no consequence to understand and characterise the geometry of the SS as that analysis is carried out in the reduced space.

Each step in the progression of transformations involves two housekeeping steps: first to locate a new origin inside the reduced subspace (referred to as *centering*) and secondly the elimination of constraints that become redundant as a result. Neither of these is computationally trivial and appropriate algorithms will be described in a subsequent chapter. As each step is applied to a progressively simpler polytope, more elaborate centering methods can be employed for the last of the reduction steps.

As described so far, choice of the origin O and basis vectors B at each step is only restricted by the requirement that they both need to fall in the reduced solution space hyperplane in order to obtain the dimension reduction, and O preferably inside the SS itself to simplify interpretation of the values vector as locating the constraint hyper-planes. But by judicious choice they can each be given additional geometric meaning.

In the most favourable scenario that all rays have been eliminated, the remaining polytope is fully bounded and it is desirable to locate the origin at its centre. The goal is somewhat ambiguous, because even for a polytope as simple as a triangle in 2D, there are many ways to define a centre point and they do not coincide except for highly symmetric cases. A reasonable criterion is the goal that any diameter should be equally bisected by the origin. As that is not possible for asymmetric shapes in general, instead the criterion can be to minimise an appropriately con-structed mean deviation from equal bisection, over all spatial directions. Even then the centre may not be uniquely defined. But it is a reasonable generalisation of the concept of a median point for a constrained 1D variable, and so the SSK origin O can be considered to define the 'most typical' feasible flux value in a similar sense. This is far more informa-tive than the single feasible flux delivered by an FBA LP calculation, which at least for Simplex is located at an extreme periphery point, namely at a vertex, and so highly untypical.

Another advantage of such a choice of origin is that in avoiding a location on the boundary hyperplanes of the SS polytope, numerical ambiguities encountered during projection to lower dimensions and which were suggested in the previous section to be the root cause of contradictory FVA results, are also avoided.

The freedom of choice for the SSK basis vectors can be exploited to encode information about its shape. Specifically, in the work to be pre-sented, a set of mutually orthogonal *maximal chords* that span the SSK is calculated, and their directions are chosen as the vector basis, which make up the rows of the matrix B. The shape and orientation of the SSK is broadly described by supplementing this with the set of actual chord lengths as calculated. To see this, consider the analogy to an ellipsoid, whose shape and orientation are specified by lengths and directions of its principal axes in space.

Making these choices, the D-specification of the SSK as in Eq. (1.12) not only gives an exact description of the SSK and SS, but also a more intuitive geometric characterisation of its location, shape and orientation.

Now consider the more common scenario where even after all avail-able projective transformations have been executed, some rays remain and so the SS is still partially unbounded.

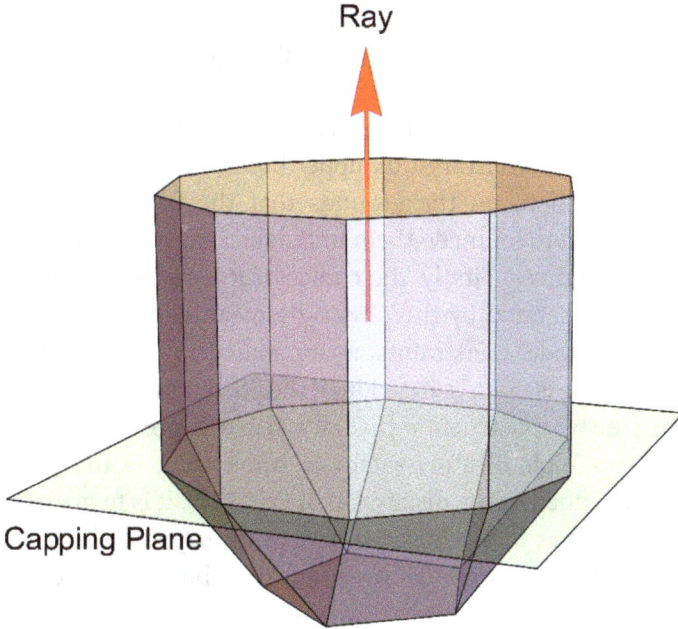

Figure 1.4 Schematic depiction of a 3D partially unbounded convex polytope SS, extending to infinity along the indicated single ray direction. The green capping plane partitions the interior into a finite section (the SSK) bordered by all its bounded facets, and the unbounded section bordered by unbounded facets.

An illustrative example is shown in Figure 1.4 that represents a facetted, semi-infinite prism in 3D. Some of its facets are bounded, and others are unbounded and extend to infinity along the single ray direction.

The shape of the prism is determined by the *bounded facets*, and they also represent the finite range of flux variation that is allowed by the explicit physical limitations on some fluxes, as well as how these are propagated by the interplay between reactions in the network.

So, it makes sense to partition the SS into the physically meaningful section bordered by bounded facets, and the unbounded part that represents a lack of knowledge about such meaningful constraints.

This is done by introducing a new capping constraint, shown as a light green plane in the figure. The ray vector is used as the constraint vector, and the distance of the *capping hyperplane* from the origin is designated as the *capping radius R*.

A value is chosen for R as the minimum value such that if it intersects any bounded facets, it is no more than tangent to them. This version of capping is called *tangent capping*. The addition of the capping constraint provides closure and the resulting bounded section of the SS becomes the SSK.

The introduction of an artificial capping constraint may seem at first to be similar to the arbitrary flux upper bounds that were discarded in the previous section, on the grounds that the failure of the second assumption listed there made them untenable. However, there are some important differences. Firstly, the choice of R is not an arbitrary value, but one that is dictated by the physically meaningful consequences of the combined model constraints, as embodied in the bounded facets. Secondly, capping is not a truncation of the SS, but merely a partitioning. It remains true by construction that all feasible SS points can be reached by adding a multiple of a ray vector to some point in the SSK, as the concept was defined in the discussion of Eq. (1.12). It is to guarantee this property, that R has to be chosen such that the capping plane is tangent at most, and does not exclude any point on a bounded facet from the SSK.

In fact, the elimination of prismatic rays as discussed for the example of a semi-infinite square column just after Eq. (1.12), can also be seen as a result of introducing a capping constraint. The capping plane defined by the vertical ray in that case is a horizontal plane. Choosing the coordinate origin in the square base of the vertical column, minimising R amounts to lowering this horizontal plane until it first intersects the single bounded facet, namely the square base. This happens for $R = 0$, at which point the capping plane coincides with the SSK. So, this is called *coincidence capping*, and differs from tangent capping in that it completely eliminates the ray direction and so allows the dimensions of the SSK to be reduced.

In the simple example of Figure 1.4, there is only one ray leading to a single capping constraint. A slightly more general situation arises if instead of a prism, the unbounded sides of the SS in the figure diverge to form a conical shape. In such a facetted cone, there is a continuum of ray directions, but an optimally chosen single capping plane might still be enough to close the SSK. However, with an awkwardly shaped facetted cone in many dimensions, it is generally found more efficient to use a small set of capping ray directions and several capping constraints need to be included in the D-specification of the SSK.

To successfully perform capping, a pre-requisite is the determination of all bounded facets of the SS polytope. So far, the concept of a facet was tacitly restricted to boundary hyperplanes of the polytope. But in fact, for example, even in 3D, the edges of a polyhedron are also facets, albeit of different dimensionality. For high-dimensional polytopes, there is a hierarchy of higher order facets and their number increases exponentially making this a major challenge. Much of the work in subsequent chapters is devoted to creating a conceptual framework and computational algorithms that achieves the goal of finding all bounded facets at all levels of the facet hierarchy, exactly for simpler cases but needing approximations for larger problems.

Once a compact SSK has been found, its location, size, shape and orientation in flux space is further analysed. A key enabler of this is the set of maximal chords previously mentioned. Their endpoints give a set of peripheral points that supplement the information supplied by the centred origin point of a typical feasible flux, and further peripheral points result from the centering procedure itself. This gives a set of flux values that define a central region of the SS, and could be viewed as a 'poor man's version' of elementary flux modes (EMs) in the sense that any linear combination of them gives a flux in this central region.

An independent though more limited indicator of the SSK location, size and central region is the maximal inscribed hypersphere that is also calculated.

Another use of the maximal chords is that they can be used to calculate the aspect ratios of the SSK when projected on various 2D planes. This reveals that typical SSKs often have quite extreme aspect ratios along some directions. This reflects the fact that some flux combinations, while not strictly constant, nevertheless have very small ranges of variation compared to others, and a further dimension reduction can sometimes be achieved by flattening out thin directions of the SSK accordingly.

1.5 Software Implementation

The aim of this book is to introduce the new concept of the FBA Solution Space Kernel, as well as a number of associated new concepts, and the computational procedure to extract and analyse it starting from the standard metabolic model specification. The description is in enough detail to allow its implementation in various programming languages.

However, its development required a specific implementation, in an environment that is powerful and flexible enough to allow experimentation and refinement, and to serve as a demonstration that the computation is indeed feasible in practice for realistically sized metabolic models. The *Mathematica* [33] programming language and notebook environment was used for this development.

The resulting software implementation allows the user to interactively take the SSK analysis through various stages of the SSK determination and analysis, for example, to make decisions on whether and how much flattening is desired. Its final outcome is both a printed report summarising key features of the SSK size and shape, and a set of data files that contain its *D*-specification, lists of ray vectors, fixed values, chord lengths, peripheral points, inscribed hypersphere and thin directions.

This implementation is freely available as a software package, under the GNU General Public License v3.0, from a GitHub repository. It may be found by searching GitHub for 'SolutionSpaceKernel' or at the URL https://github.com/wynand-verwoerd/SolutionSpaceKernel.

The repository contains all required *Mathematica* language code files in text form, as well as a detailed user manual, and both a *Mathematica* notebook to create the user interface and also alternatively script files for command line use.

References

1. B. Ø. Palsson, *Systems biology properties reconstructed networks* (Cambridge, UK: Cambridge University Press, 2006).
2. B. O. Palsson, *Systems biology constraint based reconstruction and analysis* (Cambridge, UK: Cambridge University Press, 2015).
3. J. D. Orth, I. Thiele, B. Ø. Palsson, & N. B. Author, What is flux balance analysis? *Nature Biotechnology*, **28** (2010) 245–248. https://doi.org/10.1038/nbt.1614.
4. M. W. Jeter, *Mathematical programming : an introduction to optimization* (M. Dekker, 1986).
5. M. James, The generalised inverse. *The Mathematical Gazette*, **62** (1978) 109–114. https://doi.org/10.1017/S0025557200086460.
6. C. H. Schilling, D. Letscher, & B. O. Palsson, Theory for the systemic definition of metabolic pathways and their use in interpreting metabolic function from a pathway-oriented perspective. *Journal of Theoretical Biology*, **203** (2000) 229–248. https://doi.org/10.1006/jtbi.2000.1073.
7. J. Papin, J. Stelling, N. Price, S. Klamt, S. Schuster, & B. Ø. Palsson, Comparison of network-based pathway analysis methods. *Trends in Biotechnology*, **22** (2004) 400–405. https://doi.org/10.1016/j.tibtech.2004.06.010.

8. S. Schuster, T. Dandekar, & D. A. Fell, Detection of elementary flux modes in biochemical networks: a promising tool for pathway analysis and metabolic engineering. *Trends in Biotechnology*, **17** (1999) 53–60. https://doi.org/10.1016/S0167-7799(98)01290-6.

9. S. Klamt, & J. Stelling, Combinatorial complexity of pathway analysis in metabolic networks. *Molecular Biology Reports*, **29** (2002) 233–236. https://doi.org/10.1023/A:1020390132244.

10. M. Yeung, I. Thiele, & B. O. Palsson, Estimation of the number of extreme pathways for metabolic networks. *BMC Bioinformatics*, **8** (2007) 1–15. https://doi.org/10.1186/1471-2105-8-363.

11. W. S. Verwoerd, Interactive extraction of metabolic subnets – the Netsplitter software implementation. *Journal of Molecular Engineering and Systems Biology*, **1** (2012) 2. https://doi.org/10.7243/2050-1412-1-2.

12. W. S. Verwoerd, A new computational method to split large biochemical networks into coherent subnets. *BMC Systems Biology*, **5** (2011) 25. https://doi.org/10.1186/1752-0509-5-25.

13. D. Singh, & M. J. Lercher, Network reduction methods for genome-scale metabolic models. *Cellular and Molecular Life Sciences*, 77 (2020) 481–488. https://doi.org/10.1007/s00018-019-03383-z.

14. J. D. Orth, B. Ø. Palsson, & R. M. T. Fleming, Reconstruction and use of microbial metabolic networks: the core Escherichia coli metabolic model as an educational guide. *EcoSal Plus*, 4 (2010). https://doi.org/10.1128/ecosalplus.10.2.1.

15. Z. A. King, J. Lu, A. Dräger, P. Miller, S. Federowicz, J. A. Lerman, A. Ebrahim, B. O. Palsson, & N. E. Lewis, BiGG models: a platform for integrating, standardizing and sharing genome-scale models. *Nucleic Acids Research*, **44** (2016) D515–D522. https://doi.org/10.1093/nar/gkv1049.

16. J. M. Monk, C. J. Lloyd, E. Brunk, N. Mih, A. Sastry, Z. King, R. Takeuchi, W. Nomura, Z. Zhang, H. Mori, A. M. Feist, & B. O. Palsson, iML1515, a knowledgebase that computes Escherichia coli traits. *Nature Biotechnology*, **35** (2017) 904–908. https://doi.org/10.1038/nbt.3956.

17. O. Hädicke, & S. Klamt, EColiCore2: a reference network model of the central metabolism of Escherichia coli and relationships to its genome-scale parent model. *Scientific Reports*, 7 (2017) 39647. https://doi.org/10.1038/srep39647.

18. P. Erdrich, R. Steuer, & S. Klamt, An algorithm for the reduction of genome-scale metabolic network models to meaningful core models. *BMC Systems Biology*, 9 (2015) 1–12. https://doi.org/10.1186/s12918-015-0191-x.

19. A. Röhl, & A. Bockmayr, A mixed-integer linear programming approach to the reduction of genome-scale metabolic networks. *BMC Bioinformatics*, 18 (2017). https://doi.org/10.1186/s12859-016-1412-z.

20. M. Ataman, D. F. Hernandez Gardiol, G. Fengos, & V. Hatzimanikatis, redGEM: systematic reduction and analysis of genome-scale metabolic reconstructions for development of consistent core metabolic models. *PLoS Computational Biology*, **13** (2017) e1005444. https://doi.org/10.1371/journal.pcbi.1005444.

21. A. C. Smith, F. Eyassu, J.-P. Mazat, & A. J. Robinson, MitoCore: a curated constraint-based model for simulating human central metabolism. *BMC Systems Biology*, **11** (2017) 114. https://doi.org/10.1186/s12918-017-0500-7.

22. D. J. Lugar, S. G. Mack, & G. Sriram, NetRed, an algorithm to reduce genome-scale metabolic networks and facilitate the analysis of flux predictions. *Metabolic Engineering*, **65** (2020) 207–222. https://doi.org/10.1016/j.ymben.2020.11.003.

23. A. von Kamp, & S. Klamt, Enumeration of smallest intervention strategies in genome-scale metabolic networks. *PLOS Computational Biology*, **10** (2014) e1003378. https://doi.org/10.1371/journal.pcbi.1003378.

24. A. Röhl, T. Riou, & A. Bockmayr, Computing irreversible minimal cut sets in genome-scale metabolic networks via flux cone projection. *Bioinformatics*, **35** (2018) 2618–2625. https://doi.org/10.1093/bioinformatics/bty1027.

25. C. Jungreuthmayer, G. Nair, S. Klamt, & J. Zanghellini, Comparison and improvement of algorithms for computing minimal cut sets. *BMC Bioinformatics*, **14** (2013) 1–12. https://doi.org/10.1186/1471-2105-14-318.

26. A. Röhl, & A. Bockmayr, Finding MEMo: minimum sets of elementary flux modes. **79** (2019) 1749–1777. https://doi.org/10.1101/705012.

27. M. Hossein Moteallehi-Ardakani, & S.-A. Marashi, SAFEPPP: a simple and fast method to find and analyze extreme points of a metabolic phenotypic phase plane, BioRxiv 642363, https://doi.org/10.1101/642363.

28. R. Mahadevan, & C. H. Schilling, The effects of alternate optimal solutions in constraint-based genome-scale metabolic models. *Metabolic Engineering*, **5** (2003) 264–276. https://doi.org/10.1016/j.ymben.2003.09.002.

29. N. I. of S. and Technology, DLMF: 5.19 mathematical applications. *Digital Library of Mathematical Functions*, (n.d.). https://dlmf.nist.gov/5.19#E4 (accessed March 23, 2020).

30. S. Gudmundsson, & I. Thiele, Computationally efficient flux variability analysis. *BMC Bioinformatics*, **11** (2010) 489. https://doi.org/10.1186/1471-2105-11-489.

31. R. Mahadevan, D. R. Bond, J. E. Butler, A. Esteve-Nuñez, M. V. Coppi, B. Ø. Palsson, C. H. Schilling, & D. R. Lovley, Characterization of metabolism in the Fe(III)-reducing organism Geobacter sulfurreducens by constraint-based modeling. *Applied and Environmental Microbiology*, **72** (2006) 1558–1568. https://doi.org/10.1128/AEM.72.2.1558-1568.2006.

32. J. Sun, B. Sayyar, J. E. Butler, P. Pharkya, T. R. Fahland, I. Famili, C. H. Schilling, D. R. Lovley, & R. Mahadevan, Genome-scale constraint-based modeling of Geobacter metallireducens. *BMC Systems Biology*, **3** (2009) 1–15. https://doi.org/10.1186/1752-0509-3-15.

33. Wolfram_Research & Inc., *Mathematica*, Version 12 (Champaign, Illinois: Wolfram Research, Inc., 2021).

34. K. Zhuang, M. Izallalen, P. Mouser, H. Richter, C. Risso, R. Mahadevan, & D. R. Lovley, Genome-scale dynamic modeling of the competition between Rhodoferax and Geobacter in anoxic subsurface environments. *ISME Journal*, **5** (2011) 305–316. https://doi.org/10.1038/ismej.2010.117.

35. D. Ma, L. Yang, R. M. T. Fleming, I. Thiele, B. O. Palsson, & M. A. Saunders, Reliable and efficient solution of genome-scale models of metabolism and macromolecular expression. *Scientific Reports*, **7** (2017). https://doi.org/10.1038/srep40863.

36. L. Chindelevitch, J. Trigg, A. Regev, & B. Berger, An exact arithmetic toolbox for a consistent and reproducible structural analysis of metabolic network models. *Nature Communications*, **5** (2014). https://doi.org/10.1038/ncomms5893.

37. A. Ebrahim, E. Almaas, E. Bauer, A. Bordbar, A. P. Burgard, R. L. Chang, A. Dräger, I. Famili, A. M. Feist, R. M. Fleming, S. S. Fong, V. Hatzimani-katis, M. J. Herrgård, A. Holder, M. Hucka, D. Hyduke, N. Jamshidi, S. Y. Lee, N. Le Novère, J. A. Lerman, N. E. Lewis, D. Ma, R. Mahadevan, C. Maranas, H. Nagarajan, A. Navid, J. Nielsen, L. K. Nielsen, J. Nogales, A. Noronha, C. Pal, B. O. Palsson, J. A. Papin, K. R. Patil, N. D. Price, J. L. Reed, M. Saunders, R. S. Senger, N. Sonnenschein, Y. Sun, & I. Thiele, Do genome-scale models need exact solvers or clearer standards? *Molecular Systems Biology*, **11** (2015) 831. https://doi.org/10.15252/msb.20156157.

38. L. Chindelevitch, J. Trigg, A. Regev, & B. Berger, Reply to "Do genome-scale models need exact solvers or clearer standards?" *Molecular Systems Biology*, **11** (2015) 830. https://doi.org/10.15252/msb.20156548.

Chapter 2

Hop, Skip and Jump: Finding Fixed Fluxes

2.1 Different Approaches to Identify Fixed Flux Values

The idea that a study of flux balance analysis (FBA) solutions can be simplified by a priori identification and elimination of fixed fluxes is not new to this work. For example, such an approach is advocated in Kelk et al. (2012) [1] and it suggests that Flux Variability Analysis (FVA) is employed to identify fluxes that keep the same values throughout the solution space (SS). Unfortunately, as discussed in Chapter 1, FVA is both slow and not numerically reliable to perform this task.

It seems quite plausible that just finding fixed fluxes should be a simpler project than the problem tackled by FVA to determine detailed ranges for all flux components. Indeed, there are several approaches that work well for a low-dimensional SS. However, to find an approach that scales well to large dimensions is more challenging.

For example, one might simply sample points from the SS and eliminate any flux components that differ between sample points, and retain the ones that remain fixed as a list of candidates to be further reduced. However, in high dimensions, it is not easy to adequately sample points from a polytope (more about this in a subsequent chapter) and since the total number of such frozen fluxes is unknown, it is hard to know when to terminate the sampling. In trials of this method on a problem where the number is known, a moderate sample size usually yields a list of candidates for which around 90% of the members are actual fixed fluxes. But there is a law of diminishing returns in terms of the number of new points that have to be sampled before finding a value change that allows the corresponding flux component to be eliminated from the list of candidates. So, in practice, one can never be sure that all variable fluxes have been eliminated from the list and that it is safe to terminate the sampling. This outcome is worse than obtaining an incomplete list for the purpose of dimension reduction. Eliminating an incomplete list of fixed fluxes would merely represent some missed opportunities of simplification, but a list that is too long gives an erroneous SS in which meaningful degrees of freedom have been omitted.

Another approach is to observe that the direction associated with a frozen flux is orthogonal to the SS. An example of this was shown in Figure 1.3, where the vertical axis associated with the fixed flux f_3 is perpendicular to the 2D SS located in a horizontal plane. More formally, consider two arbitrary points, X and Y, belonging to an SS that is located in a flux space of N dimensions. Writing their position vectors in terms of the flux space basis as respectively $X = (x_1, x_2, ..., x_N)$ and $Y = (y_1, y_2, ..., y_N)$, the vector $XY = X - Y$ that connects them is fully confined to the SS subspace. If component i happens to be fixed, $x_i = y_i$ and so component i of XY is zero. It follows that the corresponding flux space basis vector $f_i = (0, 0,...,1, 0,...0)$ is orthogonal to XY. As X and Y are arbitrary, this is true of any vector XY that belongs to the SS and so f_i is orthogonal to the entire SS.

It follows that if a vector basis B of the SS is known, the full set of fixed fluxes can easily be determined as the subset of flux space basis vectors that span the null space of B. The difficulty is that finding B is a large part of the end goal; it is not known at this stage. Methods based on the V-representation of the SS polytope, such as extreme pathways or elementary modes, would give B as the convex basis consisting of unit vectors pointing to all SS vertices. However, the point of the H-representation on which this study is based is to avoid finding vertices, and in any case the convex basis is both very large and vastly overcomplete, which would make finding its null space numerically intractable for a large dimensional SS.

Nevertheless, the H-representation of the SS polytope used here does allow the orthogonality requirement to be exploited to give at least a partial list. All fluxes that belong to the SS must satisfy the stoichiometry requirement, as stated in Eq. (1.1). This means that the SS is a subspace of the null space of the stoichiometry matrix S. So, all of the flux space basis vectors that are orthogonal to the null space of S, represent fixed fluxes. In practice, it turns out that about half of all fixed fluxes can be found in this way. That is helpful, as it means that just projecting these out already gives a substantial dimension reduction for subsequent steps.

But it is not the final solution, as the SS is still further subject to the objective constraint and flux range constraints which induce further fixed fluxes. In the rest of this chapter, it is shown how further elaboration that combines both the sampling and orthogonality ideas, culminate in a new algorithm dubbed the **Hop, Skip and Jump** (**HSnJ**) algorithm to find a complete list of fixed fluxes.

Doing this is facilitated by a slight rearrangement of the order in which SS constraints are implemented. The conventional approach [2] is to consider stoichiometry constraints to confine the SS to the null space of the stoichiometry matrix S. Constraints on flux ranges limit it to a finite cone in this subspace. Finally, linear programming (LP) optimisation is in effect a hyperplane that is moved across the finite cone to achieve an optimised value for a chosen objective function, and the SS is the intersection of this hyperplane with the finite cone.

In this work, however, the FBA optimisation is assumed to have been performed already. The stoichiometry constraints and the optimised objective are first applied together, to obtain a subspace termed the Objective Space (OS). The flux range constraints are applied to this afterwards eventually giving the final SS.

2.2 Orthogonal Flux Vectors: The SKIP Step

An important extension of the orthogonality approach to finding fixed fluxes is to apply it to the OS rather than just the stoichiometry matrix, and this requires determining an OS basis denoted as B_0. The flux vector G that is determined by FBA satisfies the two conditions:

$$S \cdot G = 0$$
$$O \cdot G = g \tag{2.1}$$

Here, O is a matrix containing one or more objectives as its rows, and g correspondingly is a vector of the optimised objective value or values. The OS is defined as the M-dimensional space of vectors X that satisfy Eq. (2.1). We define an objective basis for OS as the set of $M \leq N$ vectors in flux space, that when collected as the rows of a matrix B_0, allows any X to be expressed as

$$X = G + B_0^{\ t} \cdot X' \tag{2.2}$$

Here, X' is an M-dimensional vector, that in effect gives the lower dimensional representation (in OS) of the same flux that X describes in flux space. This is similar to the relationship that was illustrated in Figure 1.3 for the case $N = 3$ and $M = 2$ and G plays the role of the red arrow in the figure. B_0 is an example of the basis matrix as described in Chapter 1 and that forms part of a formal subspace specification as in Eq. (1.12).

Eq. (2.1) can be rewritten for X instead of G since all feasible fluxes satisfy it, and then substituting from Eq. (2.2), it follows that

$$S \cdot B_0^t \cdot X' = 0$$
$$O \cdot B_0^t \cdot X' = 0$$

(2.3)

Defining an extended S matrix S_{ex} by joining the rows of S and O together, it follows that Eq. (2.3) can only be satisfied for all vectors X' if

$$S_{ex} . B_0^t = 0$$

(2.4)

In other words, the basis of the OS is given as

$$B_0^t = NullSpace[S_{ex}]$$

(2.5)

For a vector X in flux space to be orthogonal to all basis vectors contained in B_0, we have $B_0 . X = 0$. To simultaneously find all flux axes that are orthogonal to B_0 in a similar way, their vectors are put as the columns of a new matrix, and that is simply the N-dimensional unit matrix. So, to extract fixed fluxes we merely multiply the $M \times N$ matrix B_0 with the N-dimensional identity matrix, and collect all columns that are zero in the product. This improves on the frozen flux list compiled before from just the stoichiometry matrix, but is still incomplete.

One final elaboration can still be used. This is to consider each frozen flux that was found as a constraint, and further extending S_{ex} by adding the fixed direction vector to its rows. Using this in an equation of the form of Eq. (2.5) yields a new basis set B_1, and this process can be iterated until it converges. In practical examples, only a few iterations are needed until the resulting list of fixed fluxes stops growing, but usually it still remains incomplete although capturing a large fraction of the frozen fluxes.

The *Geobacter* model case study discussed in Chapter 1 supplies a practical example. As shown in Table 1.1, FVA detects up to 439 frozen fluxes depending on calculation details. Merely finding flux space axes that are orthogonal to the stoichiometry matrix S as was suggested in Section 2.1, identifies 249 of these cases. This is improved by noticing that the range constraints in the model actually explicitly limits 48 fluxes to a fixed value. Extending S by adding these as additional rows, gives 306 frozen fluxes. In addition, including the objective to give B_0, increases that to 307. The iterations described previously converge quickly to the basis B_4 while fixing a total of 350 flux values. This still leaves up to 89 fixed values unaccounted for.

The iterated orthogonalisation procedure involves only straightforward matrix algebra procedures and generally executes much faster than even a single LP calculation with the same number of flux variables. It acts as a quick way to pick the low-hanging fruit, and to exploit that is followed immediately by elimination of the fixed fluxes it identified so that the remaining ones can be calculated in a reduced (lower dimension count) flux space. Somewhat whimsically it is denoted as the step that 'skips' over known frozen fluxes into the reduced subspace.

Formally executing the SKIP step requires that the constraints are projected to this reduced space, a procedure described next that is also applied subsequently in each of the dimension reduction steps still to be described.

2.3 Downcasting Constraints by Projection

One reason why the orthogonalisation is not enough to find all frozen fluxes is that the flux range constraints have not been fully taken into account yet. Although they may define independent non-zero ranges in the original flux space, after projection to the OS they can combine to induce fixed fluxes. To see that, consider the original range constraints to be given by the matrix equation

$$C \cdot X \le V \tag{2.6}$$

Some range constraints originally take the form of $\ge$ constraints. Simply multiplying them by -1 gives the desired form.

In the case of reversible reactions, it is necessary to split the range in two and construct a pair of $\le$ constraints that are equivalent to the original range. This situation will be considered in more detail later; for now, just assume for simplicity that all reactions are stated in the forward direction, so they take the form Eq. (2.6).

Taking B as the OS basis vector matrix, whether it is B_0 or its iterated version B_n, we can combine Eqs. (2.2) and (2.6) to give

$$C \cdot X = C \cdot \left(G + B^t \cdot X' \right) \le V$$
$$\therefore C \cdot B^t \cdot X' \le V - C \cdot G$$

(2.7)

From this, it follows that the equivalent constraint and values matrices in the projected space where X' is located may be taken respectively as

$$C' = C \cdot B^t \text{ and } V' = V - C \cdot G$$

(2.8)

However, to properly specify an H-representation of the SS polytope in the lower dimensional space, the rows of C' need to be normalised.

This presents a problem in any case where a row of C is orthogonal to B^t so that the corresponding row of C' is a zero vector. In such a case, however, the vector product of that row vector with X' has the same value (zero) for all points X'. The FBA point, G, is one such point and by definition satisfies this (as well as all other) constraints; so, all points X' satisfy this constraint too. That means that a constraint vector that projects to zero is redundant and its row should be eliminated from the matrix C'.

With this proviso, the prescription to find the effective constraints in the lower dimensional subspace spanned by the rows of B (downcasting) is

$$C' = \frac{C \cdot B^t}{Norm\left(C \cdot B^t \right)} \quad \text{and} \quad V' = \frac{V - C \cdot G}{Norm\left(C \cdot B^t \right)}$$

(2.9)

where *Norm* is the vector norm applied row-wise to a matrix and the vector division also understood to be applied row-wise, to the numerator expressions purged of any zero rows.

Eq. (2.9) serves as the operational definition to construct the D-specification of a lower dimensional polytope such as the OS, as given by Eq. (1.12). Since C' and V' are the constraints matrix and value vector in the lower dimensional space, they are respectively represented in Eq. (1.12) as C and V, while feasible flux G above is a particular case of the origin vector O, since by definition, it defines a point in the OS.

2.4 Sampling SS Points: The JUMP Step

Discussions are simplified by temporarily making the assumption that all flux components are positive, ranging between zero and a finite upper limit; $0 \leq X_i \leq U_i$. This corresponds to the standard formulation of LP problems. Any reversible reactions are thus assumed to have been split into a pair of oppositely directed reactions. The implications of this are further explored below. Any remaining positive lower limits L_i are temporarily removed by the replacement $X_i \rightarrow X_i - L_i$. The assumption of a finite upper limit will be removed later on.

Having found the single feasible flux G by LP, it may seem straightforward to generate further SS points by adding a constraint that excludes G. But LP applied to this just produces a new solution that differs only infinitesimally from G. So, a more subtle approach is needed to coax additional feasible points out of LP.

One such idea is to exploit the fact that LP is itself an iterative procedure that visits consecutive points in the flux space; the Simplex method in particular jumps from one SS vertex to another along a path that decreases the objective at every jump. Using S_{ext} to ensure the objective stays fixed at its optimal value, one could just intercept the jumps and record the flux vector at each. Then comparing these sample flux vectors, the components that do not change can be identified. Because the SS is a convex polytope, any interior point is a convex combination of vertex fluxes, so any component with the same value over all vertices will be frozen throughout the SS and sampling of just vertices is justified. Trials with this approach do succeed in narrowing down a shortlist of candidates, but ultimately fail because of the sampling termination problem described in Section 2.1.

A more targeted approach is to use a tailor-made objective to induce as much change in flux components as possible at each jump. For example, one could jump to a point that maximises the distance from G, while still remaining feasible. Unfortunately, the geometric distance in a vector space is the non-linear Euclidean vector norm that uses a root-mean-square combination of flux components. The resulting non-linear maximisation problem cannot be solved by simple LP.

A more explicit way to force changes in the flux components is to search for a point X in which $|X_i - G_i|$, the absolute value of the component difference, is maximised for all i. Since the absolute values are non-negative, this is equivalent to just a single maximisation of the sum of all components. This in fact amounts to maximising the so-called Manhattan distance between X and G instead of the Euclidean distance. The absolute value function used to construct the Manhattan distance is also non-linear, but there are well-known tricks to do LP problems involving Manhattan distances, at the cost of increasing the number of dimensions.

This can be illustrated by the example of a single variable in the range $0 \leq x \leq 5$, for which a feasible value x_0 has already been calculated. Suppose that it is desired to determine another feasible value, such that $|x - x_0|$ is *minimised*. This can be stated as an LP problem, by invoking a new auxiliary variable $z \geq 0$ that is connected to x by setting two additional constraints $z \geq (x - x_0)$ and $z \geq -(x - x_0)$. Together, these two constraints are equivalent to $z \geq |x - x_0|$. So, if z is minimised, it will be equal to $|x - x_0|$ and x will adjust to the value that minimises the absolute difference, within whatever feasibility constraints apply to x.

The situation is graphically illustrated in Figure 2.1 for a case with $x_0 = 3$. Minimising z in effect lowers a horizontal line across the feasible

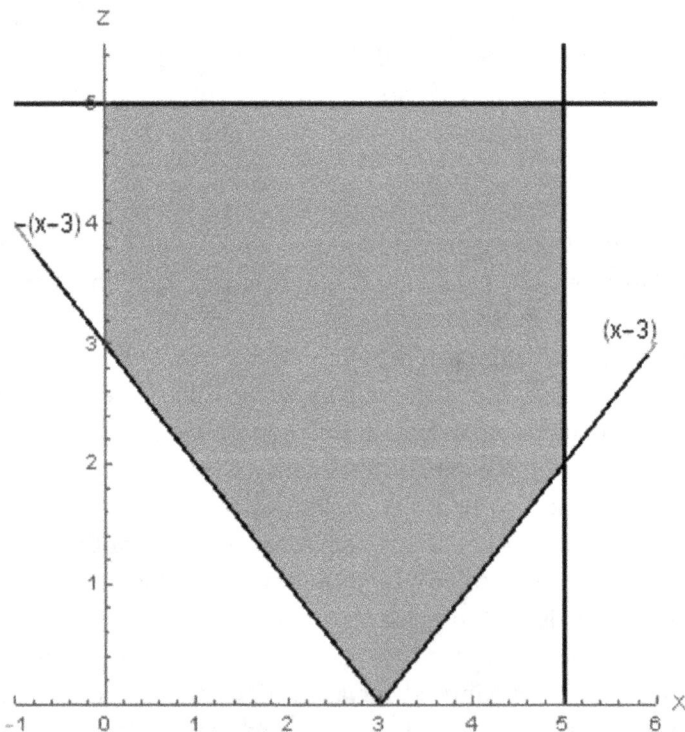

Figure 2.1   Feasible region for LP minimisation of $|x - 3|$ using auxiliary variable z.

region to find the lowest point of intersection. As shown, with no further feasibility constraints, this will select the point $(x, z) = (3, 0)$. If there are additional feasibility constraints that, for example, excludes a range of values around $x = 3$, the minimisation would yield a point on either of the sloping boundary lines, still representing $z = |x - 3|$ whichever side it is on.

This trick can easily be extended to minimise the sum of absolute values of a vector difference, by invoking an auxiliary vector Z where each of its components is subject to a pair of constraints as applied to the single variable above. But it is not directly usable for the present context because it applies to *minimisation* rather than *maximisation*.

To formulate maximisation of an absolute value as an LP problem is also possible, but requires a more elaborate extension that involves two auxiliary vectors: a vector Z as above and also a binary vector B. This is not a problem in low-dimensional spaces, but if N is already large the dimensions are tripled and, more seriously, it becomes a mixed integer linear program (MILP). This is much harder to solve and makes LP maximisation impractical for calculating fixed fluxes.

Instead, the JUMP step relies on a minor tweak of the minimisation procedure that while not maximising, still produces an x value that differs substantially from x_0. Merely changing the LP from minimising z to maximising z would give an unbounded LP problem, so in addition, z is now assigned the same fixed upper limit as applies to the corresponding x.

As is clear from the figure, maximising z for that example by use of LP just produces the trivial maximum value $z = 5$. This is of no interest, but the point is that the x value at which it finds the maximum is generally located far away from x_0. A Simplex LP solver always produces a vertex of the feasible region, so in the example, it gives either $(x, z) = (0, 5)$ or $(x, z) = (5, 5)$. If x_0 happens to be at either extreme of its range, Simplex may reproduce that value, but seems equally likely to give the opposite extreme. In more dimensions, the number of vertices that share an z-value increases so the probability of reproducing x_0 correspondingly decreases. With the Interior Point solver, it is possible that an interior x_0 value may be reproduced, but anecdotally, that only happens for one particular value (the midpoint of the range).

Accepting that it is possible but unlikely that z-maximisation reproduces the known x_0 value for one variable, when this is applied to a high-dimensional vector X, the vast majority of X-components are bound to be different.

To formalise these considerations, the JUMP step consists of solving the following LP problem. We are given the N-dimensional vector G that satisfies Eq. (2.1), more compactly written as $S_{ext} \cdot G = V = \{0, g\}$ where 0 and g respectively represent a zero vector of appropriate dimensions, and the vector of objective values. To find a new vector X, extend the space to $2N$ dimensions by constructing the vector $XZ = (X, Z)$. The components of X and Z are constrained to the same finite non-negative ranges. Then maximise the jump objective:

$$z = J \cdot XZ = \sum Z_i \qquad (2.10)$$

formed by the 'jump' objective vector $J = (0,...0, 1,...1)$, that is, it consists of N 0-entries followed by N 1-entries. XZ is subject to the set of constraints.

$$\begin{bmatrix} S_{ext} & 0 \\ -I & I \\ I & I \end{bmatrix} \cdot \begin{bmatrix} X \\ Z \end{bmatrix} \begin{matrix} = \\ \geq \\ \geq \end{matrix} \begin{bmatrix} V \\ -G \\ G \end{bmatrix} \qquad (2.11)$$

Once this has been solved to find a new X value X_J resulting from the jump, it is compared to the reference point G and all components with the same value in both are nominated as candidate frozen fluxes.

Even after the dimension reduction from projecting out orthogonal components, the dimension doubling and additional constraints means that the LP problem remains substantial for large metabolic models. Experience shows that the interior point solver is both faster and also tends to produce a larger jump, that is, it eliminates more flux components as being variable than Simplex.

A practical issue that arises in implementing the JUMP algorithm is the tolerance that is chosen for the LP problem. Relaxing the tolerance tends to induce larger variable changes, so improves the discrimination between fixed and variable fluxes and shortens the candidate list. But the price paid is that X_J can correspondingly violate the feasibility constraints by a small amount. This tendency is exacerbated by the large dimensions it needs.

When applied to the *Geobacter* model, with the fixed value tolerance criterion set at *fixtol* = 0.001, a lax tolerance of *LPtol* = 0.1 in the JUMP step identifies an additional 86 frozen fluxes (beyond the 350 found by the SKIP step) all in agreement with the ones identified by FVA and shown in Table 1.1. The more conservative setting actually used in this work is $LPtol = 0.01 \times fixtol$ (= 10^{-5} for this case) and which gives a shortlist of 93 candidate fixed fluxes. The abovementioned FVA estimates ranged between 77 and 89 additional fixed fluxes, indicating that further selection from the candidates is definitely needed in the second shortlist and possibly in the first.

One way to attempt that would be to execute further jumps starting from X_J, and possibly paring down the candidate list further. But with no guarantee that the JUMP step probes all possible flux changes, this will still not provide positive proof that the candidate list can be taken as final.

A further step is needed to obtain that certainty.

2.5 Positive Confirmation of Frozen Fluxes — The HOP Step

The goal in this section is to design a test that all candidate frozen fluxes are indeed fixed throughout the SS. A starting point is to notice that the candidate list can be divided into three groups according to the actual flux values in the reference feasible flux G.

A large first group has a value at the lower limit of the allowed flux range, the second at the upper limit and a small third group have values in between. There is some significance in this grouping further explored below, but for the present purpose, the first two groups are important because for them, the direction of any possible change in value is known. The first group denoted as 'bottoms' can only increase, and the second group called 'tops' only decrease. Suppose an LP is set up to minimise the following hop objective, subject to the feasibility constraints of Eq. (2.1):

$$H = \sum_{t \in tops} X_t - \sum_{b \in bots} X_b \tag{2.12}$$

All possible changes of any member of either group will reduce H. So, if a feasible point exists in which any of the X_t or X_b are different from their values in the reference point, minimising H will give such a point. In case the X-components are uncorrelated, the minimisation will in fact force all changeable X-components to simultaneously assume different values.

This will not happen in case there is a correlation, in the sense that, for example, for a particular pair X_i and X_j, there are feasible points such that X_i has a different value, and also points with different values of X_j, but none in which both members of the pair are different. In this situation, the minimisation of H can only adjust one of the components i or j.

To take care of such scenarios, after minimisation of H, all members of the candidate list are compared between the minimised vector X and the reference point G. Any components that differ are eliminated from the candidate list since they have been proven to be changeable. Then a new hop objective is compiled from the remaining components of G on the reduced candidate list and this is minimised in turn. This is repeated in cyclic fashion until the candidate list stops reducing, that is, no further eliminations take place.

Although this process clearly eliminates all changeable members of the tops and bottoms groups, it does not seem to address the 'in between' group of flux components. In fact, however, the members of this group are implicitly probed as well in this process. That follows because there is a direct link between changes of in-betweeners and those belonging to the tops and bottoms groups.

To explain that, it is necessary to delve into some of the mathematical background underlying LP optimisation. For concreteness, this discussion is based on the standard LP textbook by Gass [3], Chapter 3.

This discussion focusses on a *column*-oriented perspective on the constraints matrix, rather than the *row* perspective emphasised before to describe the SS geometry. In particular, Eq. (2.1) is of the general form

$$A \cdot X = B \tag{2.13}$$

Here, B is a column vector with m components, and the left-hand side of the equation is interpreted as the statement that the columns of the $m \times n$ constraint matrix A combine linearly, with coefficients given by the n components of the vector X, to give B. For now, it is assumed that the variables X_i are in the range $X_i \geq 0$ without an upper bound.

Since the LP problem is underdetermined, $m < n$ or can be made so by using row reduction to eliminate redundant rows. Then only m of the n columns of A are linearly independent. If the row reduction is taken to the extent of transforming Eq. (2.13) to reduced row echelon form (RREF), the $m \times m$ submatrix formed by these columns forms a unit matrix and the corresponding subset of m variables are termed *basic* variables. Multiple reductions are possible, each leading to a distinct subset of the columns defining the basic set. The remaining columns or variables are non-basic and are linear combinations of the basic variables. This RREF is sometimes referred to as the canonical statement of the LP problem.

A central feature of the convex polytope that represents feasible solutions to Eq. (2.13), is that each extreme point (or vertex) is associated with a particular basic set. For such a set, with the corresponding RREF reduction of A and B, the values $X_i = B_i$ for all i belonging to basic variables and $X_i = 0$ for all i that belong to non-basic variables. This follows from [3] Chapter 3, Theorem 5.

The Simplex optimisation procedure essentially consists of repeatedly modifying the basic set, by swapping a basic and non-basic variable. Geometrically, this amounts to jumping from one polytope vertex to a neighbouring one. The decision about which pair to swap at each step is made to produce a change of the objective being optimised, in the desired direction, while preserving feasibility.

For the case of flux variables in a bounded SS, the assumption that $X_i \geq 0$ needs to be replaced by $0 \leq X_i \leq U_i$. As shown in Gass [3] Chapter 9, Section 5, Theorem 1, the only change needed in the Simplex algorithm to accommodate upper limits, is that non-basic variables are allowed to be either zero or have the upper bound value. That is denoted as an extended basic solution.

Returning now to the classification of components of the reference flux G into three groups, it follows that the in-betweeners cannot be non-basic, so they must be basic variables. Every step in the Simplex algorithm to minimise the hop objective H primarily changes a non-basic variable belonging to either the tops or bottoms group, but does so by swapping it for a basic variable and so also changes an in-between variable value. Conversely, if there is a polytope vertex where a particular in-betweener (basic variable) has a different value from its value in G, the value of some non-basic variable belonging to either the tops or bottoms group also has to change and as that affects the objective H, such a change will be probed by the minimisation process.

The conclusion from this argument is that the cyclic minimisation of H that makes up the HOP step eventually probes all the flux components on the fixed flux candidates list for the possibility of change. Those that remain are a definitive list of fixed flux components.

Once this list has been determined, the projection procedure discussed in connection with the SKIP step is applied again to eliminate those frozen fluxes and reduce the problem dimensions accordingly.

The explanation above relied on the details of the Simplex LP algorithm that only samples polytope vertices. It may not be evident that it applies to the Interior Point algorithm as well, but since any variable that is fixed/variable over all vertices is also fixed/variable over the interior it is plausible that the same results should be obtained. This is borne out by trials, and has the advantage of faster computation.

2.6 Revisiting the Simplifying Assumptions

The discussion of the HSnJ algorithm so far made two simplifying assumptions that do not necessarily apply to all flux models.

First, the JUMP step required upper bounds on variables to avoid encountering an unbounded LP problem, and carrying this forward to the HOP step facilitated setting up its minimisation objective. Most current metabolic models actually satisfy this by the inclusion of artificial upper bounds on variables. However, the previous chapter advocated for removing such artificial limits. So, in implementing HSnJ it is assumed that the problem formulation may include unbounded variables, and a temporary upper limit is inserted before the JUMP step and removed after the HOP step. Its value is basically immaterial as its only function is to allow scope for variables to change their values, without becoming unrealistic. In practice, the temporary maximum is set at twice the value of the largest flux component in the reference feasible flux G, or at the default capping radius (see Figure 1.4) set by the user, whichever is the largest value.

The second simplifying assumption is that all flux values are non-negative. In the implementation this is fulfilled for models containing reversible reactions by splitting them into pairs of oppositely directed reactions before HSnJ is applied, and reunifying the reaction pairs afterwards since the rest of the SSK analysis does not require the assumption.

The splitting enlarges the flux space and introduces a spurious degree of freedom. Although this is eliminated when reconsolidating the pair, it has consequences for determining fixed fluxes, that are best explained by reference to Figure 2.2. This shows a single plane defined by the flux pair f_R and f_{-R}, for some particular reversible reaction R, out of all fluxes that make up the flux space for the network.

The diagonal line passing through the origin on the figure, represents the collection of flux points where $f_R = f_{-R}$. This is in effect a non-physical circulating flux defining the spurious degree of freedom. Similarly, for an

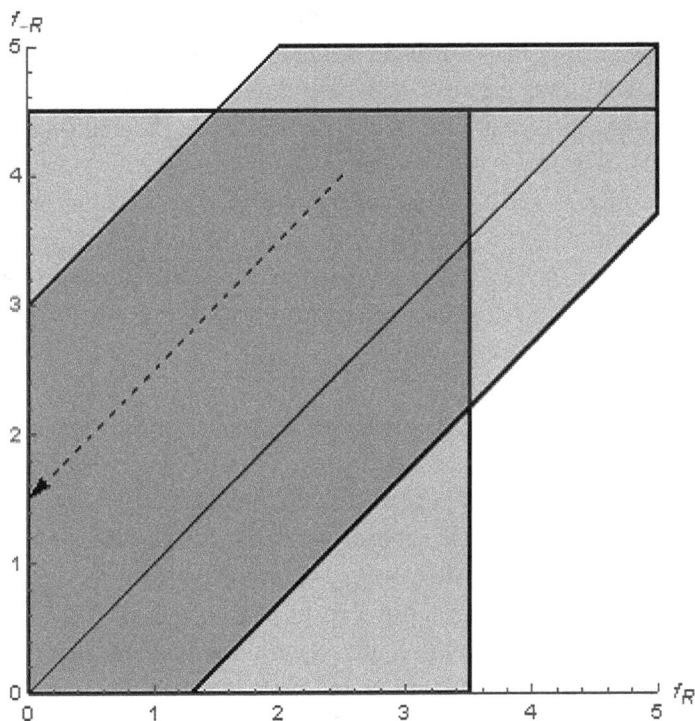

Figure 2.2 Illustrative feasible range (dark shading) for a reversible reaction separated into opposing partial fluxes f_R and f_{-R}, with upper bounds 3.5 and 4.5 respectively. Pure circulation fluxes fall on the diagonal line, while the dashed arrow shows the net flux value (on the axis) for an arbitrary feasible flux point when the circulation component is removed.

arbitrary flux point such as shown by the start of the dashed arrow on the figure, all points on the parallel diagonal line that passes through it are physically equivalent, and the meaningful flux carried by the actual reversible reaction is found from the intercept of this arrow on either axis.

For the case shown with finite upper bounds, which limit flux values to the shaded rectangle, the feasible region within this is determined by the mass flow and other constraints that result from the rest of the network containing the reversible reaction. However complex these are, in the partial flux plane the feasibility border is in general a diagonal line, since all points on such a line are equivalent regarding flux contributed to the network. There can be a single border defining a half space, or a pair that define a single diagonal band such as shown in the figure, but no more because the feasible space is convex. The case shown represents a variable flux $-3 \leq f \leq 1.3$, that is, it is fully reversible, but if the diagonal band intercepts only one axis it would indicate that a variable flux only along either the forward or backwards direction is feasible.

An extreme case of this is when the two boundaries coincide, defining a single diagonal line. In this case the consolidated flux is fixed, even though each of the component fluxes is variable.

An exception to the diagonal rule is where feasibility rules out a non-zero flux along either one of the directions. In this case, feasibility is limited to one of the two axes and diagonal lines do not come into play.

Turning now to a fixed flux calculation such as the HSnJ algorithm, there are three possible outcomes:

- Both f_R and f_{-R} are individually fixed. This can only happen when the flux point falls on an axis, so one of them will be zero and the other gives the net consolidated flux with a fixed value.
- One of f_R and f_{-R} is fixed, whereas the other is variable. Again, this can only happen for a flux point on an axis, giving a consolidated flux with a fixed direction, but a variable flux value.
- Both f_R and f_{-R} are individually variable. This corresponds to a diagonal band of either zero or non-zero width, allowing the consolidated reaction to have either a fixed or variable flux. If variable, its values may or may not uniquely determine the direction of the reaction.

In the first case, a frozen flux for a reversible reaction is successfully detected by finding frozen fluxes in the enlarged flux space. In the second case, this detects the fact that a particular reaction is constrained by feasibility to be unidirectional, although variable. In the third case, if a fixed direction or value exists, it will fail to be detected.

The conclusion is that splitting reversible reactions can give false negatives when detecting frozen fluxes, but not false positives. This is important, because excluding a falsely identified fixed flux would have erroneously reduced the SS, whereas missing a fixed flux is merely a missed opportunity at dimension reduction. Moreover, it will be seen that the SS shape analysis to be performed in Chapter 7, allows any missed fixed fluxes to be identified and eliminated at a later stage. So, this admitted shortcoming of HSnJ is more of a computational than a fundamental issue.

Whether additional effort to avoid the reaction splitting is justified, may depend on how many frozen fluxes are in fact missed. Since it takes a special situation — the coincidence of feasibility boundaries when projected to the plane of the reversed flux pair — this may be expected to be comparatively rare. That is borne out by experience with the retrospective identification of missed fixed fluxes. Also, the temporary loss of information that such misses represent is somewhat compensated by the gain of permanent information about reactions that become unidirectional. So, this issue is for now relegated to something to be addressed in future if the need arises.

2.7 Computational Performance of the HSnJ Algorithm

Although the HSnJ algorithm required considerable explanation, it is not computationally expensive. The Skip step involves straightforward matrix operations, the Jump step a single LP and the Hop step a cycle of LP calculations that usually converge in <10 iterations.

By comparison, finding fixed fluxes using FVA typically requires hundreds or even thousands of LP calculations. So, it is no surprise that HSnJ is not only more accurate, but also faster by orders of magnitude. For example, the *Geobacter* model takes more than 6000 sec to do a FVA calculation using Simplex, or around 120 sec with the Interior Point LP. The HSnJ calculation for the same data takes 0.4 sec.

Another relevant comparison is between the calculation time for finding fixed fluxes, and for subsequent stages of the SSKernel calculation. This is quite variable, since HSnJ can achieve larger reductions of the dimensions needed for subsequent stages for some models than for others. Generally, however, the time taken by the fixed flux search is negligible compared to that of later stages.

References

1. S. M. Kelk, B. G. Olivier, L. Stougie, & F. J. Bruggeman, Optimal flux spaces of genome-scale stoichiometric models are determined by a few subnetworks. *Scientific Reports*, **2** (2012) 580. https://doi.org/10.1038/srep00580.
2. B. O. Palsson, *Systems biology: Constraint-based reconstruction and analysis* (Cambridge, UK: Cambridge University Press, 2015).
3. S. I. Gass, *Linear programming: Methods and applications*, Third; Int (New York: McGraw-Hill, 1969).

Chapter *3*

Centering and Redundancy

3.1 Recurring Reduction Steps

At every dimension reduction step, of which the fixed flux removal covered in the previous chapter is the first, there are a few housekeeping steps that have to be performed. The first of these, the downwards projection of points in flux space and of constraints, was described in Chapter 2 and formally defined by Eqs (1.12) and (2.9). To perform this projection, it is necessary to find a new origin O located in the lower dimensional hyperplane, as illustrated in Figure 1.3. Subsequent work is facilitated by further requiring that O is located inside the lower dimensional solution space (SS) polytope.

In fact, it is desirable that it should be well into the interior, as far as possible from the boundaries. Recalling that in the H-representation of Eq. (1.3), the elements of the values vector V denote orthogonal distances from the origin to the boundary hyperplanes, this will ensure that these elements are all significantly positive. If a subset of these values is zero or near zero, that means that the origin lies on the intersection of the corresponding subset of boundary hyperplanes. In that case, minor numeric inaccuracies can easily produce negative values, placing the origin outside the polytope. From this perspective, the Flux Balance Analysis (FBA) solution G is not a very suitable origin point, particularly if found by Simplex, which places it at a vertex and so at the intersection of a maximal number of boundary hyperplanes.

These considerations establish *centering* as a recurrent housekeeping task to be performed at each dimension reduction step. The goal is to find a 'centre point' for a polytope in the half-space specification, that is, defined by a linear constraint set.

Unless the polytope is highly symmetric, no unique centre can be expected. Even for the simplest two-dimensional (2D) polytope, a triangle, there are several classical definitions of a centre point and the list is growing; the online 'Encyclopedia of Triangle Centers' lists tens of thousands of proposed definitions. A well-known general-purpose centre is the Chebyshev centre, defined according to preference as the centre of either the minimal circumscribed ball that includes a given polytope, or alternatively its maximal inscribed ball. The first version is not usable here because it may fall outside the polytope. The second does have its uses, but as easily imagined for an elongated rectangle, is not unique and may deviate far from a position that is optimally equidistant to all boundaries. It is further explored in Section 3.2.3.

A few pragmatic approaches for finding a point that is 'reasonably' central to a high-dimensional polytope, are discussed below. Multiple approaches are needed because there is a trade-off between effectiveness and scalability to high dimensions, as well as the ability to deal with the partially unbounded polytopes encountered in earlier stages.

Once the downwards projection has been done, it also becomes necessary to tidy up the redundant constraints it can produce. Any constraints with defining vectors orthogonal to the reduced space are automatically eliminated during the projection. But others can remain and need dedicated procedures to find them. That is the subject of the recurrent *redundancy elimination* task that is described in Section 3.3.

3.2 Centering Algorithms

For a *V*-specified polytope, a good interior centre point can in principle be found simply as the centroid of all vertices. Although this works in low dimensions, it does not scale well because of the exponential increase of the vertex count with dimensions. For example, a hypercube in a relatively modest $N = 30$ dimensions already has more than 10^9 vertices placing it at the upper limit of what is practical for the centroid calculation, not to mention finding the vertices first.

Using the *H*-specification instead, it delivers one point on each boundary hyperplane without calculation (the orthogonal intersection from the origin) and taking the centroid of these would seem to be a viable alternative. For the 30-hypercube there are 60 such points so are easily calculated. However, depending on the location of the current origin, some orthogonal intersections may well fall outside the polytope if it is irregular or has a large aspect ratio. One remedy that sometimes helps is an iteration loop that moves the origin to the new centre and recalculates the orthogonal intersection points. But this is not guaranteed to converge to an interior point and so a more elaborate method is needed.

3.2.1 *The Constrained Linear Distance Algorithm*

For a polytope specified by $C \cdot x \leq V$, the orthogonal distances to the constraint hyperplanes are given by the elements of V. At a good centre point, these elements will all be positive numbers of similar magnitude. For a highly regular polytope, they might all be equal; but when the aspect ratios are different in different directions, the best that may be achievable is to have them in reasonably balanced pairs for opposite directions.

The norm of the vector V can be interpreted as a combined measure of how far the current coordinate origin is from all sides of the polytope. If the origin is far outside the polytope, this norm will be large and it will be decreased as the origin is moved closer. So, a plausible working hypothesis is that moving the origin to minimise the norm will give a reasonably central interior point.

Suppose that the origin is shifted to g, the old and new coordinates are related by $x = x' + g$ so the new inequalities become $C \cdot x' \leq V - C \cdot g = V'$. The desired g that minimises the norm of V', amounts to the best approximate solution to the equation $V - C \cdot g = 0$. Except for the case of a simple cone, an exact solution is ruled out because that would imply that all constraint hyperplanes intersect in a single point. The best approximate solution is the g that minimises the vector norm $\|V - C \cdot g\|$, whether this is the L1 (absolute value) or L2 (Euclidean) norm.

To minimise the Euclidean norm is particularly simple to program, as the properties of the Moore-Penrose pseudoinverse introduced in connection with Eq. (1.5) guarantees that the least squares solution is

$$g = C^+ \cdot V \tag{3.1}$$

This method is dubbed the Least Squares Distance (LSD) centering. In fact, it turns out that LSD gives the same point as that found by the iteration loop based on the orthogonal intersection centroid. It has the same drawback too, that g is not always a feasible point inside the polytope, contrary to the working hypothesis.

A counterexample is the trivial one-dimensional (1D) constraint pair $\{x \le 2, x \le 5\}$ where the obvious solution is $g = Centroid(2, 5) = 3.5$, which falls outside the H-specified polytope $x \le 2$. This example is somewhat artificial, as the second constraint is actually redundant, but examples without redundant constraints can easily be constructed in two or three dimensions, and also the constraint set is in any case usually not free of redundant members until after being made so by the centering process.

An alternative is to minimise the sum of *absolute* values of $(V - C{\cdot}g)$. As outlined in the discussion of Figure 2.1, this can be done using linear programming (LP). In itself that does not solve the problem, but the advantage is that the polytope constraints can be added to this LP so that the solution becomes guaranteed not to be exterior, restoring the working hypothesis.

This algorithm is described as the Constrained Linear Distance (CLD) algorithm, and its mathematical formulation is as follows.

First, the set of m inequalities is converted to equalities by introducing a set of slack variables $s_i = V_i - (C \cdot x)_i$ with $i = 1, m$. Constraining $s_i \ge 0$ ensures that the resulting point x will be interior. Then, a second set of m auxiliary variables z_i are introduced to minimise absolute values, such that $z_i \ge V_i - (C \cdot x)_i$ and $z_i \ge -(V_i - (C \cdot x)_i)$ which implies that $z_i \ge |V_i - (C \cdot x)_i|$. The objective is to minimise $\sum z_i$.

In matrix form, the LP constraint equations are

$$
\begin{bmatrix} I_m & 0 & C \\ 0 & I_m & -C \\ 0 & I_m & C \end{bmatrix} \cdot \begin{bmatrix} s \\ z \\ x \end{bmatrix} \begin{matrix} = \\ \ge \\ \ge \end{matrix} \begin{bmatrix} V \\ -V \\ V \end{bmatrix}
\tag{3.2}
$$

Here, I_m is the m-dimensional identity matrix and the optimisation is carried out in a $(2m + n)$ dimensional vector space with an objective vector $O = \{0,..0,1,..1,0,..0\}$. It is in fact possible to streamline this formulation because with the s_i constrained to be positive, they reduce to $s_i = |V_i - (C \cdot x)_i|$ and so we can dispense with the z-vector and simply minimise $\sum s_i$ and so reduce the dimensions to $(m + n)$.

With this simplification, the CLD procedure scales quite well and remains fast to calculate even in several hundred dimensions. By construction, the CLD centre is always feasible, but turns out to generally be located on the boundary. In high dimensions, it is in fact usually on the intersection of several boundary hyperplanes (i.e., many of the elements of V' are zero), which is not ideal.

The reason for this behaviour is that because of the linear behaviour of the L1-norm, gains in moving towards one boundary tend to balance the losses of moving away from the ones opposite. That is in contrast to the L2-norm that applies in the LSD method, where the quadratic increments penalise increased distances more severely. So, in CLD it becomes acceptable to make many distances zero at the expense of increasing others.

On the other hand, both LSD and CLD algorithms have the advantage that they can be applied to open as well as closed polytopes. They only involve minimising the combined intersection distance to whichever constraint hyperplanes are present, and are unaffected if there are directions (rays) in which the polytope is unbounded.

3.2.2 *The Constrained Linear Ordered Differences Procedure*

To avoid boundary points during centering, CLD can be modified to minimise the sum of *differences* between distances from the origin to the orthogonal intersection points on all boundary hyperplanes, rather than the distances themselves.

This idea is quite straightforward to put in matrix form by defining a differencing matrix D, as illustrated by the following 4D example:

$$D_4 \cdot X = \begin{bmatrix} 1 & -1 & 0 & 0 \\ 1 & 0 & -1 & 0 \\ 1 & 0 & 0 & -1 \\ 0 & 1 & -1 & 0 \\ 0 & 1 & 0 & -1 \\ 0 & 0 & 1 & -1 \end{bmatrix} \cdot \begin{bmatrix} x_1 \\ x_2 \\ x_3 \\ x_4 \end{bmatrix} = \begin{bmatrix} x_1 - x_2 \\ x_1 - x_3 \\ x_1 - x_4 \\ x_2 - x_3 \\ x_2 - x_4 \\ x_3 - x_4 \end{bmatrix} \tag{3.3}$$

For n dimensions, a pseudocode definition of the differencing matrix is:

$$
\begin{aligned}
&\textit{For row} = 1 \textit{ to } \tfrac{1}{2}n(n+1) \\
&\quad \textit{For } i = 1 \textit{ to } n \\
&\quad\quad \textit{For } j = i+1 \textit{ to } n \\
&\quad\quad\quad D_n[row, i] = 1 \\
&\quad\quad\quad D_n[row, j] = -1 \\
&\quad\quad \textit{Endfor} \\
&\quad \textit{Endfor} \\
&\textit{Endfor}
\end{aligned}
\tag{3.4}
$$

Applying D to both sides of the constraint equation gives the appropriate set of constraints for the set of intersection distance differences. The same reasoning that led to Eq. (3.2) applied to this gives

$$
\begin{bmatrix} I_m & 0 & C \\ 0 & I_m & -D_m \cdot C \\ 0 & I_m & D_m \cdot C \end{bmatrix} \cdot \begin{bmatrix} s \\ z \\ x \end{bmatrix} \begin{matrix} = \\ \geq \\ \geq \end{matrix} \begin{bmatrix} V \\ -D_m \cdot V \\ D_m \cdot V \end{bmatrix} \tag{3.5}
$$

Although the structure of the equations is very similar, the variables s and z now represent different things and so the CLD simplification to omit z is no longer available. Calculation of a centre based on Eq. (3.5) gives quite good results for low-dimensional polytopes. Not only is it less prone to give boundary points than CLD, but it comes much closer to the ideal of well-balanced distance values.

However, the price to pay is a large increase in matrix dimensions because the number of differences grows quadratically with the number of variables. This makes it impractical for the number of dimensions typically encountered in FBA models.

Several tweaks can be used to deal with this problem.

First, notice that it is not really necessary to include all differences because they are not independent; for example, in Eq. (3.3), $(x_2 - x_3) = (x_1 - x_3) - (x_1 - x_2)$. If the differences are assigned to a matrix $\Delta_{ij} = (x_i - x_j)$, the ones displayed in Eq. (3.3) define the upper triangle of Δ. It can be seen that only the entries on the first subdiagonal $(\Delta_{i,i+1}, i = 1, n - 1)$ are independent. If the rest of the upper triangle differences are expressed in terms of them, the sum of all differences reduces to a weighted sum of subdiagonal differences. The quantity of interest is the sum of absolute values of differences rather than the differences themselves, but the triangle inequality can be used to show that the weighted sum of absolute subdiagonal differences gives an upper limit for the upper triangle sum of absolute values. So, minimising the weighted sum of absolute differences still tends to reduce all differences, albeit with reduced efficiency for the ones beyond the subdiagonal. But in exchange there is a vast reduction in the matrix size, with the growth restored to a linear increase with n.

To implement this, Eq. (3.5) is still used, but the definition of the differencing matrix D is changed to produce only subdiagonal elements:

$$\begin{aligned}
&For\ \ row = 1\ to\ n-1 \\
&\quad D_n[row, row] = 1; \\
&\quad D_n[row, row+1] = -1; \\
&Endfor
\end{aligned} \qquad (3.6)$$

Also, the objective is changed to a vector of the subdiagonal *weights*, applied to the z-components.

The second tweak is based on the observation that the weights in the objective turn out to follow a parabolic curve, with the maximum weight allocated to the middle entries. This can be exploited by making sure that these entries represent the longest 'diameters' of the polytope. To achieve that, the constraints are first reordered into pairs with their constraint vectors as nearly opposite to each other as possible. Then, the pair with the largest sum of individual elements in the values vector V is placed near the middle position (row = ½ n) and the remaining pairs on either side of this, and so on, until all positions are allocated. This reordering process is the reason the resulting Constrained Linear Ordered Differences (CLOD) method is described as using ordered differences.

In trials, CLOD is found to be considerably slower than CLD, but still viable even for fairly large metabolic models. It retains the advantage of being applicable to open polytopes. It is less prone to produce a centre on the boundary, and when it does, the actual number of hyperplanes that intersect at the centre is much less than for CLD.

For these reasons, CLOD is the preferred method for polytopes that are partially unbounded, except for very large dimensional cases where CLD may be more practical.

3.2.3 *The Maximal Inscribed Hypersphere Centre*

Once all ray directions have been eliminated either by projection or by tangential capping, it becomes possible to construct a finite hypersphere that has the maximal possible radius while still remaining inside the feasible solution space. The centre position of this inscribed hypersphere (the *inscribed centre*) may not be unique, but as long as the radius is finite, this guarantees that the centre is inside and away from all boundaries by at least this distance. Conversely, if the radius is zero, this gives positive confirmation that some boundary hyperplanes coincide, that is, there is at least one flux direction with a fixed flux value.

The calculation of the inscribed hypersphere is once more based on linear optimisation. As before, consider a polytope specified by the constraint equations $C \cdot x \leq V$. In general, the v_i entries in the values vector V could have any values, but in the special case that the coordinate origin is well inside the polytope (i.e., not located on any boundary hyperplane) the entries $v_i > 0$. Geometrically each one represents the perpendicular distance to constraint hyperplane i, so a hypersphere with radius $r = v_i$ will be tangent to this hyperplane at the perpendicular intersection. For all values $r \leq v_i$ the hypersphere will be fully contained in the feasible half-space determined by constraint i. It follows that choosing r to be:

$$r = \underset{i=1,m}{Min}[v_i] \qquad (3.7)$$

ensures that the hypersphere centred on the origin, will simultaneously satisfy all feasibility conditions, and be fully inscribed in the polytope.

As seen before, vector V changes when the origin is moved, so the radius obtained from Eq. (3.7) changes accordingly. The goal pursued here is to find the origin that produces the largest inscribed hypersphere, and take that as the definition of the polytope centre point.

An optimisation such as this that maximises a minimum, often referred to as a minimax problem, is readily transformed to the standard LP formulation by introducing auxiliary variables. As in Section 3.2.1, the origin shift to a position g gives the new values vector $V' = V - C \cdot g$. Defining a new vector $z = (Z, Z, Z, ...Z)$ with m equal components, we add the set of constraints $z \leq V'$. These are equivalent to the inequality $Z \leq Min[v_i]$, and which becomes an equality if Z is maximised. Substituting V' the new constraints reduce to $C \cdot g + z \leq V$.

To ensure that the new origin remains interior, we also require $C \cdot g \leq V$. The objective is to maximise the single value Z, but to cast that in the required form of an LP objective vector in the extended space of vectors $\{x, z\}$, the objective to be maximised is of the form $(0, 0, 0, 1, 1,...1)$. Finally, since the target vector in this space is freely adjusted by the LP solver, an additional set of constraints is needed to ensure that all components of z are kept equal during optimisation.

In matrix form, the combined constraint set is given by

$$\begin{bmatrix} C & 0 \\ C & I_m \\ 0 & B \end{bmatrix} \cdot \begin{bmatrix} x \\ z \end{bmatrix} \begin{matrix} \leq \\ \leq \\ = \end{matrix} \begin{bmatrix} V \\ V \\ 0 \end{bmatrix} \quad \text{where} \quad B = \begin{bmatrix} 1 & 0 & 0 & ... & -1 \\ 0 & 1 & 0 & ... & -1 \\ \ddots & \ddots & \ddots & \ddots & -1 \end{bmatrix} \qquad (3.8)$$

Here, B is an $(m-1, m)$ matrix acting to set all z_i to a common value. The x value that maximises the stated objective, is the sought for minimax centre, g.

The formulation above will normally succeed in eliminating all zeroes in the values vector associated with an original origin located on the polytope boundary. In this case, it also delivers the maximal inscribed sphere radius as the optimised Z value, which gives a first indication of the shape of the SS polytope. More specifically, this radius is a maximal lower limit on the length of any chords that span the polytope, a fact that will become useful in a later section dealing with chord calculations.

But there is one caveat. If the radius comes out as zero, it means that there is some direction in flux space where two boundary hyperplanes coincide. That will happen, for example, if a fixed flux failed to be eliminated by the Hop, Skip and Jump (HSnJ) procedure of Chapter 2, and as shown there the presence of reversible reactions in the metabolic model can result in such a failure. In this case, it is unavoidable that g will fall on the boundary and the corresponding pair of v_i will be zero. It might be hoped that even so all remaining zeroes will be eliminated. Unfortunately, that is not the case; all solutions with one or more zeroes are equivalent as far as maximising Z is concerned, so LP can return any of them. The implication is that the centre may in this case not be moved to the interior to the full extent that is in principle possible.

This approach is also not guaranteed to find the best (that is, the most symmetrically located) inscribed hypersphere. For example, in 2D, in an elongated rectangle, there is a range of equivalent positions for an inscribed circle along the centre line, and anecdotally LP seems to produce a position at either end of the rectangle. Even worse, if the long sides of the rectangle are made to diverge slightly to form an elongated trapezium, the maximal inscribed sphere is unequivocally located at the thick end even though there are arguably more central points available.

With the high aspect ratios observed for many metabolic solution spaces, this may become a real issue. The strong point of the inscribed centre is that it is well away from boundaries, not necessarily that it is optimally centered.

3.2.4 *Iterative Centre Refinement*

The previously discussed centering methods perform well enough for their purpose in the intermediate stages of the kernel space reduction. However, in the final stage, it becomes a primary goal that the coordinate origin is chosen at a point that is as central as possible, so that it can be used as a typical, even representative, point for all feasible solutions.

For this task, a special refinement procedure was developed. It assumes that the current origin is a feasible point, preferably already well inside the polytope. Then, it samples a selection of directions chosen to give a good coverage of all directions in the high-dimensional hyper-angle space. For each direction, the offcut radii to the boundary intersections, both forward and backwards from the origin, are found. The origin is then shifted to the centroid of all intersection points for all sampled directions. Any difference between the forward and backwards radii for each direction tends to be reduced in this way. Sampling of directions is repeated at the new origin, and this is iterated until convergence is obtained.

The final result is a point that lies near the centre of each separate diameter defined by the sampled directions, and also defines the centroid of all the periphery points where the diameters intersect the boundary.

Perfect pairwise equality of the offcut radii is only possible for a hypersphere or a sufficiently symmetric polytope, so a measure of the deviation from this ideal gives some indication of the polytope shape at least regarding symmetry. A quantitative measure called 'centrality' is introduced below to serve as criterion to terminate the iterations, as well as an overall shape indicator.

For this programme to work, calculation of the intersection radii needs to be quick even in high dimensions. The key to this is that only straightforward trigonometry is involved. Given a constraint unit vector $\hat{c}$ with its value vector entry v and a chosen direction $\hat{s}$, the cut-off radius along $\hat{s}$ is $v/\hat{c}\cdot\hat{s}$, provided that $\hat{c}\cdot\hat{s} > 0$. For each of the forward and backwards directions of $\hat{s}$, only a subset of constraints will satisfy this provision. Calculating the cut-off distance for all members of each subset, the minimum value of each subset gives the required forward and backward cut-off radii to the polytope boundary, r_f and r_b, and which combine to give the diameter of the polytope.

The next issue is the selection of sampling directions. Three distinct sets of directions are combined to form the sampling set:

- Directions from the origin to a predetermined set of fixed points on the polytope periphery. In each iteration, as the origin shifts, these directions change.

The fixed points used are the endpoints of the maximal chords, as outlined in Chapter 1, Section 4, in the discussion of the basis vectors used in the D-specification of Eq. (1.12), and detailed in Chapter 7.

- The constraint vectors defining the polytope boundary hyperplanes.
- A uniformly spaced set that is independent of the polytope, namely a set of directions that specify the vertices of a regular n-dimensional simplex. The orientation of the simplex is randomly chosen in each iteration cycle to sample additional directions and avoid any bias.

By definition, the simplex vertices are equally spaced over the appropriate hyperangle space. For example, in 2D these would be the three 120 deg separated vertices of an equilateral triangle; in 3D the four 109.47 deg separated vertices of a regular tetrahedron, and in general in n dimensions, there are $(n + 1)$ directions with a uniform angle separation $\text{ArcCos}[-1/n]$, that approaches 90 deg for large dimensions.

A quantitative criterion is needed to establish how well centred the refined centre is. For a given direction, the centre of the corresponding diameter line is located at $0.5(r_f + r_b)$. Taking r_f as the larger of the two radii, the discrepancy between this point and the one delivered by the refining algorithm is $r_f - 0.5(r_f + r_b) = 0.5(r_f - r_b)$. Expressing this as a fraction of the diameter and taking the mean over all sampled directions yields the centrality deviation as a percentage by the expression

$$cd = 50\,\text{Mean}\left[\frac{\left|r_f - r_b\right|}{\left(r_f + r_b\right)}\right] \qquad (3.9)$$

Perfect centering would give $cd = 0\%$, and a value better than 10% is usually achieved for a low-dimensional Solution Space Kernel (SSK). The cd value is partly a measure of the success of the centering strategy, but the degree of regularity of the polytope sets a lower limit to cd so it also gives an indication of polytope shape. Values tend to increase in larger dimensions but values around 20% are usually achievable.

The centre refining iteration is judged to be converged if either the centrality changes by <0.05%, or the centre position shifts by <2% of the mean of all offcut radii. Large n models sometimes reach a stage where minor fluctuating shifts of the centre no longer yield significant improvements, so the iterations are also terminated if the iteration count exceeds the larger of 25 and $(n/4)$.

As described, a polytope specified by m constraints in n dimensions will usually yield $(2m + n + 1)$ directions and $2(2\,m + n + 1)$ intersection points on the polytope periphery. The exact number can vary somewhat because directions that are near duplicates can arise from the distinct sets, and are eliminated, and additional directions are added in some circumstances as described below.

A secondary outcome of the centre refining is the set of periphery points on the boundary of the polytope used in the last centering iteration. These periphery points are considered to be a valuable extension of the information about the polytope and its shape, beyond just knowledge of its centre. By construction, they define the extension of the polytope in a reasonably homogeneously spaced set of directions. Their limited number means that it is a far less detailed specification than that afforded by the full, exponentially large set of vertex points. On the other hand, this makes it also far more manageable, and arguably gives a sampling of a central region of the polytope. Moreover, convexity implies that any convex linear combination of the periphery points gives another feasible point. So, the periphery points can be considered to define the vertices of an inscribed polytope that fully defines such a central region. This polytope is referred to as the *peripheral point polytope* or PPP in subsequent work.

The procedure as described still needs to be protected from one potential pitfall. If the initial coordinate origin is at or near an intersection of several constraint hyperplanes (e.g., a polytope vertex) it may happen that many of the sampling directions do not intersect the boundary hyperplanes at all except at the origin, and those that do only intersect it very nearby. In this case the centroid of diameter intersection points may only differ insignificantly from the starting point, that is, the centre 'remains trapped in the corner'. What is needed in this situation is to ensure that there is at least one sampling direction that aims across to the other side of the polytope, roughly bisecting the interior angle formed by the nearby boundary hyperplanes.

It is in fact possible to find such a direction by taking just the offending hyperplanes (identified by their zero or near zero entries in the values vector V) to construct an unbounded polytope. The desired direction will be a ray of this auxiliary polytope. In the next chapter, various algorithms for finding rays are described, and one of those, designed to find capping rays, is particularly well suited as it produces rays that tend to bisect interior angles.

So, the refining procedure includes a corner trapping avoidance check, after the initial selection of sampling directions in each iteration, to test whether there are boundary hyperplanes nearby to the current origin. If so, the appropriate capping ray directions are added to the sample.

The need for such a protection is particularly acute if the preliminary centering was done with the CLD or CLOD algorithms, as they are prone to produce a centre at multiple hyperplane intersections. This need is largely eliminated if the inscribed hypersphere centering is used before attempting the refinement. But whenever it finds a zero inscribed radius, the possibility of a peripheral centre exists and the refinement procedure with inclusion of corner trapping avoidance is helpful to move the centre to an interior position.

3.2.5 *Illustrative Examples*

A comparison of how the various centering methods perform on 2D polygons is useful to highlight some of the features mentioned in the earlier discussion, and is shown in Figure 3.1. The example polygons are variously defined by constraints relative to coordinate origins inside, outside and on the periphery, and with shapes ranging from the highly regular octagon to some elongated ones with moderately high aspect ratios.

It is observed that the simplest method, LSD, gives good centering except in cases where it fails completely by delivering an exterior (infeasible) point. This is prevented by CLD that enforces feasibility, but at the price of always yielding boundary points, mostly vertices. The Constrained Ordered Linear Differences (CLOD) method improves on this and its centre point seems visually often well centred, although in the cases where LSD points are external the CLOD point tends to be close to the boundary but still feasible. The inscribed sphere is always well interior but can migrate towards one end of elongated polygons. It nevertheless serves as a good starting point for centre refinement, which universally delivers a point symmetrically situated close to the centre of a series of diameters that pass through it and define the periphery point pairs that are also shown on the figure.

For most cases, the centres remain distinct, although for the regular octagon, all methods except CLD converge to the 'actual' centre.

Finally, the figure shows how the central region polygon that the periphery points define, fit into each constraint polygon. They typically cover 70% to 80% of the polygon surface area, and reflect its shape in a general way while ignoring the extremes represented by vertices. For illustrative purposes, the refining shown actually did not include the

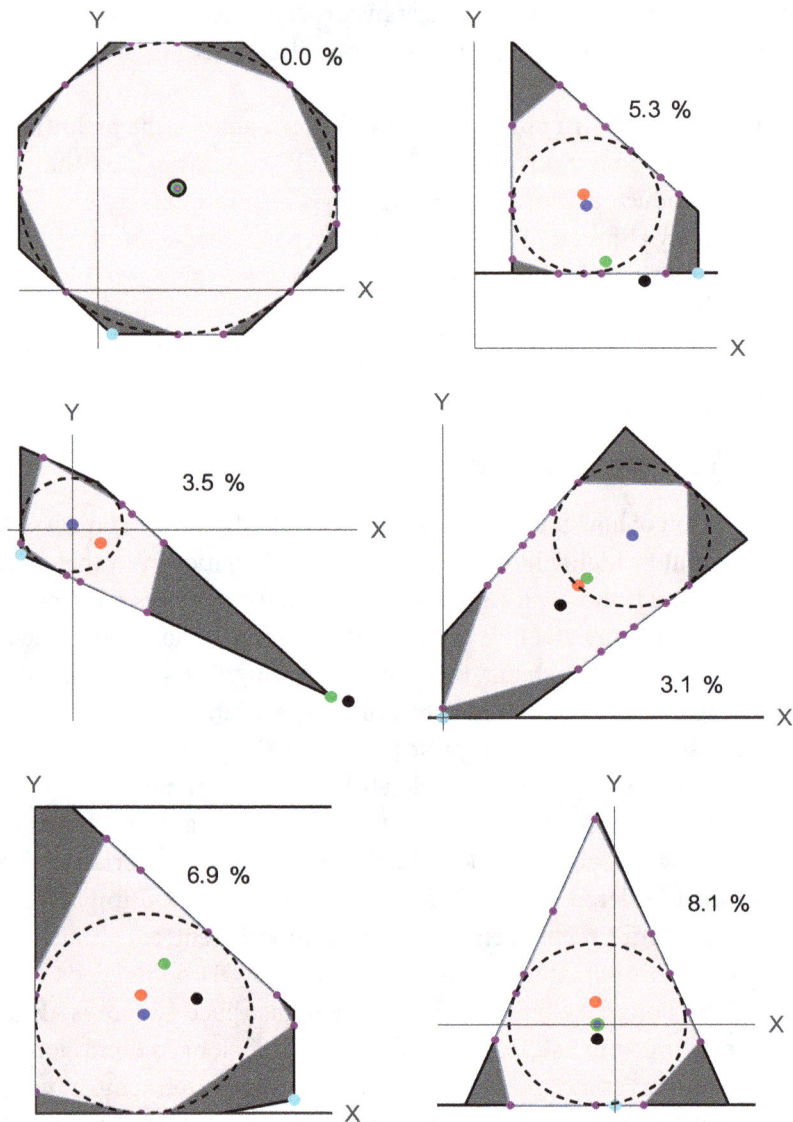

Figure 3.1   Polygons (shaded in grey) and their calculated centres from various algorithms shown by coloured dots: Black — LSD; Cyan — CLD; Green — CLOD; Blue — Inscribed circle; Red — Refined centre. Purple dots are periphery points from centre refinement and the pink polygon is the central feasible region (the PPP) that they define. The centrality deviation *cd* of the refined centre is shown as a percentage next to each figure.

maximal chord endpoints used in larger dimensional examples. These would by definition include three or four vertices and give almost perfect coverage in 2D, not representative of results achieved in multidimensional cases.

The 2D examples cannot reflect the full complexity of such multidimensional polytopes for which the algorithms are designed. Still, the remarks above do convey a sense of the observations made on dealing with high-dimensional SSK cases encountered in metabolic models. The standard inclusion of maximal chords means that the principal shape-determining vertices are included in the periphery points. So, the central polytope reflects the overall shape somewhat better in multidimensional polytopes than for the 2D illustrations.

Note that for the 2D examples, the inscribed circle area is a rough indication of polytope area. That statement does not hold for multidimensional cases. With this understood, the inscribed radius can nevertheless be taken as a rough measure of SSK spatial extension even for multidimensional cases.

A minor tweak made on the grounds of the observations, is that LSD and CLD are best used in combination. Whenever CLOD is too slow, LSD is tried first and only if it fails to give a feasible point is CLD used to avoid this. The inscribed hypersphere is used as soon as the SSK becomes closed, for example, by tangent capping, and the final centering is done using the refinement procedure.

3.3 Redundancy Elimination

The central focus in constructing the SSK, is the stepwise reduction of dimensions. But in the process, the number of active constraints is also reduced systematically. One way this happens is that a constraint vector that is orthogonal to the projection hyperplane H_p, describes a boundary hyperplane that does not intersect H_p and so becomes irrelevant. Such cases are already eliminated during the projection as described in Eq. (2.9).

However, additional constraints may become redundant after projection, for example, if multiple projected constraint vectors become identical, even if with different constraint values. This case is described as collinear constraints.

It is worth noting that the concept of redundancy is a little different for equations and inequalities. For equations, constraint equations that are linearly dependent are redundant and are eliminated using matrix row reduction as was done for example during the fixed flux HOP step discussed after Eq. (2.13). For inequalities, they can be redundant even if linearly independent. A trivial example is the 1D set $\{x \leq 2, x \leq 5\}$ in which every x that satisfies the first inequality, also satisfies the second linearly independent constraint which hence is redundant.

An algorithm to find all such constraints in the H-representation of a polytope will be presented below, but requires an LP for each constraint and can be computationally challenging for large dimensions. Hence, a simpler method that just removes collinear constraints is discussed first.

3.3.1 *Collinear Constraint Removal*

For this method, the individual vector overlaps between all constraint vectors are calculated as a single matrix multiplication $C \cdot C^t$, where (as before) C is the matrix that contains all the normalised constraint vectors as its rows. Row i in the upper triangle (excluding the diagonal) of the product matrix gives all overlaps of constraint vector i with constraints $j > i$ and if any of those equal 1, they are collinear with i. From this set of collinear vectors, only the one with the smallest entry in the values vector is retained since it makes the others redundant similarly to the earlier 1D example. Repeating this process for all i completes the redundancy removal.

This calculation is very fast and surprisingly effective at removing most redundant constraints that arise after projection.

Complete removal of redundant constraints is not strictly necessary, since their presence does not in principle invalidate the H-specification of a polytope. However, their presence may cause numerical problems, such as spuriously shifting a calculated polytope centre to a suboptimal position. So, applying collinear removal is a worthwhile step at the early dimension reduction stages in large metabolic models, and is also useful as a pre-processing step in the more elaborate method discussed next because it reduces the number of constraints that need to be tested for redundancy.

3.3.2 *Explicit Redundancy Testing by LP*

Any constraint hyperplane that can be omitted from C without changing the feasibility polytope, is classified as a redundant constraint. A common scenario is that the hyperplane passes through a vertex already defined by the rest of the constraint set (referred to as the *remainder polytope*). Another case is where the hyperplane does not even intersect the remainder polytope at all. In the converse case, if the hyperplane does intersect the remainder polytope and not just at a single vertex, it forms part of the boundary and so it is not redundant because omitting it produces a different polytope.

To test constraint i for redundancy, an LP calculation is performed based on the standard polytope constraint set $C \cdot x \le V$ but with three modifications:

- A small increment δ is temporarily added to the V_i entry
- Constraint inequality i is temporarily changed to an equality
- A trivial objective, namely the zero vector, is used.

The first modification provides for the case that hyperplane i passes through a vertex. Since the origin is interior, the increase in the offcut value displaces the hyperplane outwards so it does not intersect the polytope any more. To achieve this despite possible numerical inaccuracies, δ is chosen larger than these. On the other hand, it should be small enough to prevent inadvertently removing a hyperplane that significantly intersects the remainder polytope. A value $\delta = 0.01 - 0.1$ is usually suitable.

The second modification ensures that any feasible point must lie *on* hyperplane i, not on either side of it. So, if this hyperplane does not intersect the remainder polytope, the LP will have no feasible solutions and this signals that constraint i is redundant.

The trivial objective indicated as the third modification is just a computational convenience, since only the existence of a feasible point is relevant, not the objective value.

All constraints are tested in this way, one by one, in a loop. If the feasibility test succeeds, the constraint is retained and reverted to its original form before moving on to the next. If it fails, the constraint is immediately removed and the next test applied to the reduced constraint set. This ensures that where two constraints make each other redundant without both being redundant, only the first one encountered is removed.

Since redundancy only becomes an issue after fixed fluxes and prismatic rays have been removed (see next chapter), the flux dimensions encountered in redundancy removal usually only count in the hundreds and the process outlined is reasonably efficient. This is further improved by applying collinear removal first before entering the testing loop.

Chapter 4

Finding Rays

4.1 Ray Vectors, Ray Space and Convex Bases

In the preview of Chapter 1, Section 4, three major tasks were outlined. Having covered the first in Chapter 2, the spotlight now turns on the second task of finding directions in which an unbounded polytope extends to infinity. This was used as a working definition of the concept of a ray, and graphically illustrated in Figure 1.4.

In the literature on unbounded polytopes, unbounded directions are sometimes referred to as *recession directions* and a prominent concept is that of *extreme rays.* An extreme ray is defined as a ray that cannot be written as a non-trivial convex combination of other rays, and geometrically the extreme rays define the cone edges for the example of a multidimensional cone. There is a close analogy between extreme rays and extreme pathways as used in metabolic models, and this relationship as well as methods for computing extreme rays is further explored by Gagneur and Klamt (2004) [1]. The computation of extreme rays is subject to the problems of a combinatorial explosion in their numbers, just as was discussed for polytope vertices and elementary modes in Chapter 1.

For the same reasons that the vertex representation of the flux polytope is avoided, the approach presented here focusses just on *rays*, not extreme rays. The aim is to find a limited number, of the order of the dimension count, and use these in the work that follows.

To put this in context, the concept of the **ray space** of a polytope is introduced. The ray space contains all the polytope rays and is a subspace of the vector space in which the polytope is embedded, but not all vectors in the ray space are necessarily rays. A simple example is shown in Figure 1.4 where the vertical axis forms the one-dimensional (1D) ray space (a subspace of the three-dimensional (3D) space that contains the polytope). Note that the upwards direction defines the only ray in this case, but the downwards direction, while also a vector in this space, is not a ray.

The goal set here is to find a set of ray vectors that span the ray space, possibly as an overcomplete basis.

In general, any convex linear combination of ray vectors is also a ray. Since the set of extreme rays by definition collects all rays that cannot be expressed as a convex combination of other rays, it follows logically that any ray can be written as a convex combination of extreme rays. So, the set of extreme rays forms a *convex complete basis* of the set of all polytope rays. This is a stricter requirement than is put here on the desired set of rays, which merely forms a *vector complete basis* of the ray space.

The comparison is analogous to that encountered to describe a closed convex polytope. Any point in the interior of such a polytope can be represented in terms of any set of N basis vectors that span the N-dimensional space of the polytope. But this basis is not unique, and also there are vector combinations that are not interior points. By contrast, the set of all vertex vectors of the polytope forms a convex basis. This means that all interior points can be written as convex combinations of vertex vectors, and all such combinations are interior points. The convex basis is obviously unique, but the number of vectors that belong to it is larger than the space dimensions and can become unmanageably large for multidimensional polytopes.

Similarly, for an unbounded polytope, the extreme rays are unique, and form a convex basis, but their number tends to increase exponentially with dimensions. On the other hand, the ray basis to be constructed here is not unique, and only vectorially complete. Its number of members has to be at least the dimensionality of the ray space, but is meant not to be very much larger. The non-uniqueness and flexibility about the exact member count will be exploited to attempt maximising the range of rays that can be expressed as convex combinations of this ray basis. Not being able to guarantee the inclusion of all rays is the price paid for keeping the ray basis tractable.

This is deemed acceptable firstly because rays are considered to be of only peripheral interest when interpreting the biochemical meaning of the solution space polytope (which is the reasoning for excluding rays from the Solution Space Kernel (SSK) in the first place). Second, the convergence of the ray finding process to be described and the validation tests performed later for specific models suggest that the vast majority of rays are in fact included. For simple cases, the ray basis found this way in fact turns out to be convex complete anyway.

The first step in this endeavour is to formulate the mathematical definition of a ray vector. Given an N-dimensional polytope specified by the feasibility constraint set $C \cdot f \leq V$ where C is a $M \times N$ (constraint) matrix, and assuming an interior coordinate origin that guarantees

$v_i \geq 0$, a ray vector is defined as a unit vector $\hat{x}$ along a direction x that satisfies

$$C \cdot x \leq 0 \qquad\qquad (4.1)$$

This is indeed a ray vector, because the vector $a\hat{x}$ where a is an arbitrary large positive scalar $a > 0$, will still satisfy Eq. (4.1) and therefore also the feasibility constraint set. That means that the point $a\hat{x}$ remains in the interior of the polytope no matter how far from the origin, so $\hat{x}$ is an open direction.

The remainder of this chapter explores different cases and strategies to find solutions to Eq. (4.1).

4.2 The Reduced Solution Space: Linealities and Prismatic Rays

Consider a special case of Eq. (4.1), when the inequality is replaced by a strict equality. This case is solved by the standard procedures of linear algebra, by any vector x that belongs to the null space of matrix C. So, whenever NullSpace[C] is not empty, the orthonormal basis vectors of this subspace are a set of ray vectors that are called **linealities**. If x is a lineality, $a\hat{x}$ satisfies both the ray condition (4.1) and the feasibility condition irrespective of the sign of a, that is, arbitrarily large vectors along both $\hat{x}$ and $-\hat{x}$ remain interior. Thus, a lineality can be seen as a bidirectional ray.

The simplest biochemical example of this is a reversible reaction for which the rest of the network places no restriction on its flux, and it can form a virtual loop running either way.

Actual metabolic networks usually do not contain linealities, but if they are present, they are quick and easy to determine as described and eliminate by projection. So, without loss of generality consideration can now be limited to cases where the ray x is such that at least one component of the vector $C \cdot x$ is strictly negative.

Also, removal of lineality directions means that all vectors that are orthogonal to C (namely its null space) have been projected out, and the remaining space will thus be spanned by the constraint vectors (the rows of C). This fact will be used in later work.

A second group of rays can also be easily found just by a matrix operation on C. If there is a constraint vector c_i that is orthogonal to all other constraint vectors, the vector $x = -c_i$ satisfies Eq. (4.1) because the element i of $C \cdot x$ is -1 and all other elements are zero. This may seem an unlikely situation, so Figure 4.1 illustrates an example.

The H-specification of the octagonal prism shown takes the form

$$
\begin{bmatrix}
0 & 0 & -1 \\
1 & 0 & 0 \\
0.707 & 0.707 & 0 \\
0 & 1 & 0 \\
-0.707 & 0.707 & 0 \\
-1 & 0 & 0 \\
-0.707 & -0.707 & 0 \\
0 & -1 & 0 \\
0.707 & -0.707 & 0
\end{bmatrix}
\cdot
\begin{pmatrix} x \\ y \\ z \end{pmatrix}
\leq
\begin{pmatrix} 0.5 \\ 2 \\ 2 \\ 2 \\ 2 \\ 2 \\ 2 \\ 2 \\ 2 \end{pmatrix}
\tag{4.2}
$$

It is clear from both the picture and the explicit matrix form that the first constraint vector, representing the left-hand boundary plane, is orthogonal to all of the remaining eight constraint vectors that specify the octagonal sides. The same situation obviously applies for a prism with a cross-section of any polygonal shape, and it can be generalised to N dimensions by defining a semi-infinite prism as the result of erecting orthogonal sides in the N-th dimension on any polytope in $(N - 1)$ dimensions.

The negative of such a constraint vector, in this case $\hat{z}$, is an example of the type of ray designated from here on as a **prismatic ray**.

All such rays are easily found by calculating the overlaps between all constraint vectors as the elements of the matrix product $C \cdot C^t$. If a row of the overlap matrix is zero except for a diagonal 1, the corresponding constraint vector is orthogonal to all others and its negative is a prismatic ray.

Removal of a prismatic ray can conceptually be seen as introducing a capping hyperplane defined by the ray vector, and adjusting its capping radius to make it coincide with the single constraint plane used to define it. In the case of Figure 4.1, the plane perpendicular to the ray would be parallel to the XY-plane shown in green, and when moved to coincide with the 'bottom' of the prism on the left, it in effect removes the Z-direction completely allowing both of the coinciding constraint planes to be removed. In practice, explicit construction of a capping plane can

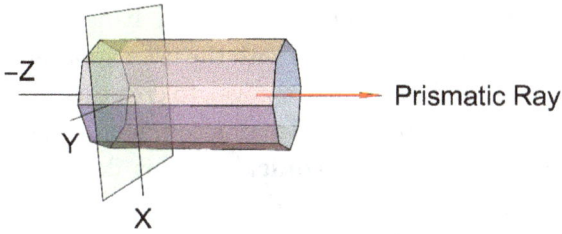

Figure 4.1 Prismatic ray for an SS polytope shaped as a semi-infinite octagonal prism stretching to infinity along the positive Z-axis.

be omitted and it is enough to merely move the origin to the defining boundary hyperplane, drop the corresponding constraint row from C and make an orthogonal projection of all other row vectors to the reduced flux space with the ray direction removed. For the illustrated 3D example that leaves a 2D octagon as a bounded SSK. In multidimensional cases, some ray directions usually survive this coincidence capping and have to be further dealt with.

A tricky point that arises is the necessity for the origin to be at least marginally interior. If it should happen that any v_i is exactly zero, that is, the origin is mathematically exactly on the boundary, the direction of the constraint vector c_i becomes ambiguous in the sense that c_i and $-c_i$ will define equivalent constraints. Then, a prismatic ray direction cannot be assigned uniquely. Or from a different perspective both directions become rays, that is, the prismatic ray cannot be distinguished from a lineality. It was to avoid such problems that emphasis was placed in Chapter 3 on the importance of centering procedures to avoid the boundary wherever possible.

It turns out that for most metabolic models, there are a large number of prismatic rays, often the majority of all rays. For example, for Geobacter, it turns out that there are 257 rays in total, of which 144 are prismatic rays. This may seem surprising, but the reason is that having disposed of stoichiometry constraints in projecting the SS to the objective hyperplane, the remaining constraints originate from range constraints on the fluxes. Since the constraint vectors of these align (positively or negatively) with the axes of the full flux space, they originally belong to two orthogonal sets. The projections involved in eliminating fixed fluxes and transforming to the objective hyperplane destroys some of this orthogonality, but whatever remains account for the prismatic rays that are found.

Linealities and prismatic rays are the low hanging fruit and it is expedient to combine their removal with the prior application of stoichiometry constraints and fixed flux removal, to create a Reduced Solution Space (RSS) in which the more challenging computation and elimination of the remaining rays, termed **conical rays**, is done. The Geobacter model, for example, starts from 2586 constraints on 940 flux variables, and yields an RSS with 138 constraints on 109 variables.

4.3 The Ray Matrix

The general case of the inequality, Eq. (4.1), can be converted to a set of equalities by defining an auxiliary set of slack variables s_i, one for each constraint. The resulting matrix equation in the enlarged space formed by joining the M-component slack vector s to the N-component flux vector x is

$$\dot{I}_M \; C \cdot \begin{pmatrix} s \\ x \end{pmatrix} = \qquad\qquad (4.3)$$

To preserve the less/equal condition of Eq. (4.1), the s-vector has to be mildly positive (M-positive), that is, $s_i \geq 0$ with at least one value strictly positive. The customary methods of linear algebra do not allow for enforcing the M-positive condition, especially on just a subset of the variables. The remedy is to once more convert this to an optimisation problem to be solved using linear programming (LP) methods.

This requires that an objective is defined so it can be optimised, and so creates an opportunity to guide ray construction towards a basis with desired properties, such as a maximal coverage of the ray space.

A first step is to note that row i of Eq. (4.3) reduces to

$$s_i = -c_i \cdot x \qquad\qquad (4.4)$$

In other words, the slack variable values in a given (s, x) solution vector calculated, for example, by LP, can each be interpreted as the negative of the overlap between vector x and the corresponding (normalised) constraint vector. Because Eq. (4.3) is homogeneous, the normalisation of (s, x) is arbitrary. The ray vector merely indicates a direction in flux space so should be a unit vector. So, after calculating it, (s, x) is normalised such that its x part on its own has a norm value of 1. It follows that then $0 \leq s_i \leq 1$, because each value is the overlap between two unit vectors, and geometrically all the overlaps are negative because otherwise x would intercept the constraint hyperplane and would not be a ray.

Solutions to Eq. (4.3) as delivered for example by LP automatically include the s-components, but Eq. (4.4) shows that even if a ray vector x is found in some other way, the s_i can be calculated from knowledge of the constraint vectors and are unique. It is convenient to retain the full (s, x) vector that represents both the actual ray and its relationship to the constraint vectors, as a unit in further analysis. In fact, as LP solutions by nature only delivers one ray at a time, it is useful to collect the 'extended' ray vectors (s, x) together, as the rows in a progressively growing matrix, called the **ray matrix**.

The $(M + N)$ columns of this matrix separates into two distinct groups: the first M columns contain all the slack values, one for each constraint, and the last N columns the actual ray vectors in flux space.

At any point in the ray search, the ray matrix denotes the current, possibly still incomplete ray basis. It is incomplete in the sense that only a limited range of rays can be formed by convex superposition of the rays in its row vectors. At the start, it may even be vectorially incomplete, meaning that the ray basis does not yet span the ray space even in terms of unrestricted linear combinations.

In building the ray matrix, the first goal is to make the ray basis vectorially complete, and thereafter to extend its convex coverage to include the widest range of rays achievable.

It is in principle possible to construct a ray matrix that is guaranteed to be convex complete: namely the ray matrix R composed of all extreme rays. But based on estimates of elementary mode numbers mentioned in Chapter 1, such a ray matrix might easily have a row count of many millions.

So, the goal here is to achieve the best coverage possible with a manageable ray matrix. To be vectorially complete, the ray matrix must have at least as many rows as the ray space dimensions, and it is found that a suitably constructed ray finding method can substantially converge with the number of rows at a low multiple of the dimensions. For a low-dimensional ray space, it may indeed be possible to achieve full convex completeness without breaking the bank; for example, the 3D polytopes of Figures 1.4 and 4.1 each has a 1D ray space and the ray matrix only needs a single row.

A typical example of a metabolic model ray matrix is shown in Figure 4.2 as a colour plot representing element values. Columns 1 to 81

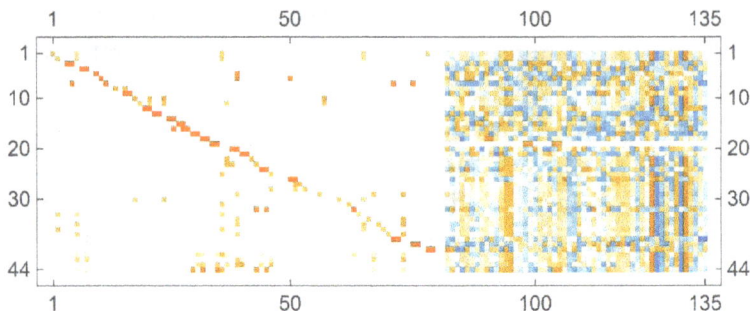

Figure 4.2   Ray matrix for a metabolic model after reduction to 81 constraints on 54 flux variables. There are 44 rays, spanning a 44-dimensional ray space. Each numerical element is plotted as a coloured square, positive values in reddish shades, zero in white and negative values in bluish shades.

show slack variable values, and 82 to 135 show the actual normalised ray vectors. The slack section is noticeably a sparse matrix. That is no coincidence.

Equation (4.4) can be used to find the slack elements of a ray formed by a convex combination of known basis rays. Let (s, x) and (σ, y) be two rows of the ray matrix. Then, with $z = a\,x + (1 - a)\,y$ and $0 \leq a \leq 1$, the new ray vector is $\hat{z} = z/|z|$ and its i-th slack entry is

$$-c_i \cdot \hat{z} = -c_i \cdot (a\,x + (1-a)y)/|z| = [a\,s_i + (1-a)\sigma_i]/|z| \qquad (4.5)$$

Since all numbers on the right-hand side are non-negative, this means that a ray with a zero entry in slack column i can only be formed by convex superposition of ray matrix rows with zeroes in the same column. On the other hand, a non-zero entry can be formed by a combination of zero and non-zero entries.

In geometric terms, when a ray has a zero entry in slack column i, it is orthogonal to c_i, and so it is aligned with constraint hyperplane i because all vectors in this hyperplane are by definition orthogonal to c_i. So, the algebraic observation above reflects the fact that an arbitrary ray that is aligned to a hyperplane requires basis vectors that are similarly aligned in order to be expressible in that basis.

An (s, x) vector with multiple zero slack entries may either be aligned with the single intersection of all the corresponding hyperplanes (extreme rays would be an example of that) or with multiple subsets of them. For example, the single ray in Figure 4.1 aligns with the eight different pairwise intersections of unbounded boundary planes of the octagonal prism.

The conclusion from either the algebraic or geometric standpoints, is that to represent the widest range of ray vectors, basis vectors appearing in the ray matrix should be selected to satisfy two criteria:

- Each individual row should have as many zero *s*-values as possible so that it can contribute to rays aligned with as many constraint planes as possible. Geometrically, this can be interpreted as constructing basis vectors that are as *peripheral* as possible.
- Non-zero slack entries should be distributed as uniformly as possible over all columns when taking the ray matrix as a whole.

The following sections discuss how this strategy is used in practice to construct a ray matrix efficiently. As will become clear, the ray matrix contains crucial information that allows characterisation of various polytope properties such as for classifying facets. It will play an important role in much of the work presented in subsequent chapters.

4.4 Singleton Rays

Singleton rays are a slight generalisation of prismatic rays. A singleton ray direction only overlaps (negatively) with a single constraint vector, although unlike a prismatic ray it is not antiparallel to it. These can be calculated from a simplified version of Eq. (4.3), namely

$$[U_k\, C] \cdot \begin{pmatrix} s \\ x \end{pmatrix} = 0 \qquad (4.6)$$

where U_k is the vector that forms the *k*-th column of the identity matrix (i.e., its *k*-th element is one and all others zero). Also in Eq. (4.6), *s* represents a single slack variable, rather than a vector of slack variables as in Eq. (4.3). Once Eq. (4.6) is solved, the *s*-value it yields is inserted as s_k to form the full extended ray vector, all other slack values being set to zero. The single non-zero slack value is the origin of the term **singleton ray**.

Equation (4.6) is solved by standard matrix algebra as

$$\begin{pmatrix} s \\ x \end{pmatrix} = NullSpace[U_k\, C] \qquad (4.7)$$

As stated before, this procedure allows no control over the arithmetic signs of the solution vector components, but because there is only a single s-value, this is easily remedied retrospectively by simply reversing the (s, x) vector in case s comes out as negative.

Equation (4.7) may yield multiple rays with the same s_k, in the case that the null space is multidimensional. However, to find all the singleton rays, Eq. (4.6) needs to be repeatedly solved in a loop that lets k run over all M constraints. This computation is considerably more onerous than finding prismatic rays, which only requires a single matrix multiplication, but is still quicker than repeated LP calculations.

An example of singleton rays in a 3D polytope is shown in Figure 4.3.

As shown, there are two singleton rays, one aligned with each of the slanted rectangular sides, and falling in the XY-plane. Each is orthogonal to the constraint vector that defines its aligned face, as well as the constraint vectors along $\pm\hat{z}$ that define the triangular faces, and so only overlaps with the constraint vector of the opposing slanted side.

To demonstrate that prismatic rays are a special case of singleton rays, notice that for the case where the top angle of the 3D wedge is 90 deg, this overlap would also be zero and the rays shown would become prismatic rays.

In principle, it might be possible to extend this approach to find ray vectors that only overlap with, for example, a pair of constraints vectors. But not only does the number of required LP solutions increase combinatorially in such an extension, it also becomes very challenging if not

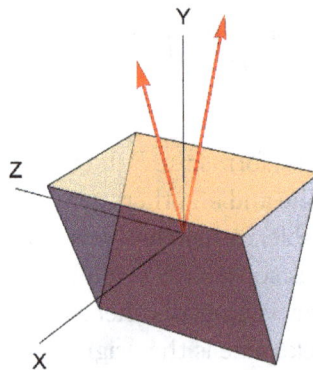

Figure 4.3 Example of singleton rays (red arrows) for a wedge-shaped unbounded 3D polytope. The polytope is open at the top and has four sides extending to infinity in the Y direction, two triangular and two rectangular.

impossible to maintain the M-positive requirement on the slack values and this is not pursued any further.

Metabolic models typically yield far fewer singleton rays than prismatic rays, and because they are not orthogonal to all constraint vectors, they cannot in general be projected out as simply as was the case for prismatic rays. Instead, once calculated, the singleton rays are used to construct a starting ray matrix, which is then further extended by the LP-based peripheral ray calculation described in section 4.5.

4.5 Peripheral and Aligned Rays

The calculation of peripheral rays is based on the goals set in Section 4.3 for an efficient ray matrix. The first of these is that each individual extended ray vector should have as many as possible zero s-entries. It would be possible to set this up as an LP optimisation involving binary indicator variables, but that leads to a mixed integer LP that becomes intractable in large models.

Instead, the strategy followed here is to simply minimise the sum of s-entries. This is (perhaps surprisingly) found very effective, as evidenced by the large number of zero entries in the slack section of the ray matrix calculated in this way for the example of Figure 4.2. Several factors plausibly contribute to this success. Because all slack entries for a given ray are non-negative and relatively independent of each other, a large number of non-zero values tend to add up to a large sum. Also, a convex combination of peripheral rays generally has more non-zero slack entries than the rays that constitute it. In addition, the triangle inequality implies that the norm of such a combination is smaller than one, further increasing the sum when the combination is properly normalised. These trends mean that rays that are the most peripheral tend to have both a small number of non-zero entries and a small value for their sum.

The LP implementation seems straightforward by combining the constraint set formulated in Eq. (4.3) with the (s, x)-space objective vector $\{1, 1, \dots 1, 0, 0, \dots 0\}$ and minimising the resulting sum $S = \Sigma s_i$. However, the pitfall is that this will merely produce the trivial zero vector solution.

That is prevented by adding a single constraint that fixes a chosen slack value s_k to an arbitrary finite value, say 0.5. There is no loss of generality, because the LP solution vector (s, x) is at any rate normalised retrospectively such that its x-part is a unit ray vector, and this fixes s_k to its proper value.

Occasionally, the LP fails as infeasible, and that is positive proof that no ray exists that has $s_k > 0$. This applies to all rays, not just the subset known at this stage of ray matrix construction. Such a matrix column is described as *consistently zero*.

So, by probing each column k in turn in this way, multiple new peripheral rays are added to the ray matrix while at the same time gleaning information about any consistently zero columns in the s-section of the matrix.

This can go wrong if, for example, the LP solver returns a solution with a large x-vector length, say a value of 10^6. In this case, normalising (s, x) as described reduces s_k to the value 10^{-6}. This would be indistinguishable within numerical tolerance from a zero entry, which contradicts the starting assumption of a non-zero value. Such ambiguities or inaccuracies are eliminated by only using a Simplex LP solver for the ray search. Any of the centering methods described in Chapter 3 ensures that the origin is in the vicinity of the bounded facets and so the Simplex method, that by design delivers a vertex point, always produces x-vectors of reasonable length. The same is not true of the Interior Point method for an open polytope.

The strategy of systematically probing each k-value in turn has the advantage of distributing non-zero entries over all columns of the ray matrix, another of the goals set in Section 4.3. However, in its simple form, it leads to duplication. Setting s_k to 0.5, the solution returned is either a duplicate singleton ray or also contains other non-zero entries. In the latter case, when those columns are probed in turn, they will usually produce the same ray as when column k was probed, that is, further duplicates.

A better strategy is to maintain a pool of *probe columns* that are due to be checked and is dynamically updated after each LP test. At any point in the search, this contains only columns that are still zero in the current ray matrix, but excluding any that have been proven by a feasibility failure to be consistently zero for all rays. So, the set of probe columns reduces at each iteration, and the cycle ends when there are no probe columns left. This strategy guarantees that after the probing cycle, any column has no more than one non-zero entry, so all rays constructed this way are linearly independent. That is helpful to the further goal set before, of constructing a vectorially complete basis. But it also risks missing some rays, as there is no reason in principle why the sets of non-zero slacks of any two rays cannot overlap.

A further refinement was tried in which after the first probe of column k, each non-zero slack column of the resulting solution is constrained to a zero value in turn and the probe repeated. This does deliver additional rays, but at a low strike rate in terms of the number of additional LP calculations involved. So, this idea was abandoned in favour of the more targeted finding of aligned rays as described below.

The use of a set of probe columns can further be exploited to construct **facet aligned** rays that are aligned with specific facets of the polytope. A facet lies in the intersection of one or more boundary hyperplanes, and is orthogonal to all of the constraint vectors of those intersecting hyperplanes. So, a ray aligned with the facet has zero overlap with all such constraint vectors. This means that setting slacks to zero for all such columns and excluding them from the probe column list, produces rays aligned with the chosen facet hyperplane.

The combination of the singleton ray calculation and the column probing, exhaustively tests for rays that overlaps with any constraint vector. Recalling that, once linealities have been eliminated, the set of constraint vectors span the polytope space, it follows that if the ray collection is still empty after that it proves that no rays exist for the polytope. In this case, the ray search can be terminated and the conclusion drawn that the polytope is closed.

Having found a set of rays as described so far, the next step is to establish if they form a complete vectorial basis for the ray space. But as remarked before, the ray space may be only a subspace of the polytope space, and of unknown dimensions so far. Fortunately, the set of consistently zero columns after probing determines that conclusively.

Each consistently zero column identifies a constraint vector that is orthogonal to all rays, including those not represented in the ray matrix, and is denoted as a **rayfree** constraint vector. Collectively, the set of consistently zero columns identify a basis for the rayfree subspace, which is the complement of the ray space. So, the ray space dimension is the difference between the ranks of matrix C and its submatrix C_{rf} consisting of the rayfree constraint vector rows.

Using this yardstick, it is usually found that the ray matrix remains incomplete after the probing round. To remedy that, a vector basis is first constructed for the 'missing' part of the ray space not yet spanned by the current collection of rays in the ray matrix, and designated as a target set. The target vectors are defined as orthogonal to both the rayfree subspace and all known ray vectors. Then rays are found that have non-zero overlaps with these target vectors, because that will expand the reach of the ray matrix into the missing part. These are denoted as **aligned rays**, but note that they are (partially) aligned with the target directions, not with constraint hyperplanes as was the case for **facet aligned rays** mentioned earlier.

Specifying the steps formally, first the target vectors are the rows of a matrix T given by

$$T = \text{NullSpace} \begin{bmatrix} C_{rf} \\ X \end{bmatrix} \tag{4.8}$$

where X is the ray section of the current ray matrix. Next, the constraint matrix is augmented by adding the target vectors as additional constraints and the ray matrix acquire corresponding additional slack columns:

$$\tilde{C} = \begin{bmatrix} C \\ T \end{bmatrix}; \quad \tilde{R} = \begin{bmatrix} S_C & S_T & X \end{bmatrix} \tag{4.9}$$

The probing cycle is then repeated with the augmented matrix $\tilde{C}$ in an enlarged vector space (s_C, s_T, x). The objective vector has unit entries for both sets of slack columns, but probing is restricted to only the s_T columns associated with T. At its conclusion each column of the additional slack matrix S_T will have non-zero entries, so there is guaranteed to be a ray that overlaps with each target vector. Finally, deleting S_T from $\tilde{R}$ yields the appropriate set of rows to be appended to the previous vectorially incomplete ray matrix.

Although the target vectors do form an orthogonal basis for the missing ray subspace and there is a ray overlapping with each of them, there is no guarantee that the aligned rays will form a complete basis for it. As a counterexample, in 3D, it is possible to take three vectors in the plane $x + y + z = 0$; each of these vectors individually overlaps with all three of the basis vectors along the coordinate axes, but together they still only span a plane. Nevertheless, in many trials on different metabolic models a single application of Eqs. (4.8) and (4.9) has always proven sufficient to extend the ray matrix to a vectorially complete set. If it should ever fail — as easily established by checking the ray matrix rank against the known ray space dimension — the procedure can be applied again using the new ray matrix as a starting point.

Summarising the discussion, constructing a vectorially complete ray matrix consists of three main steps:

- Singleton rays are found by matrix algebra operations to supply an initial ray matrix.
- LP-based column probing of the ray matrix extends it with additional rays, establishes the rayfree subspace and ray space dimensionality.
- Extending the ray matrix further by targeted aligning of rays with missing directions, gives a complete ray basis.

The ray matrix R constructed from peripheral rays proves to be an essential tool in finding and analysing polytope facets in chapters 5 and 6. Moreover, it turns out that those procedures can occasionally yield still further rays that are added to the ray matrix. But the rarity with which that happens, suggests that the peripheral ray procedure already gives a very adequate characterisation of the ray space even if it is not convex complete.

An example of the outcome of the ray matrix calculation was shown in Figure 4.2. In this example, there were no singleton rays, seen in the figure as no rows with only a single entry in the slack section. There are, on the other hand, a number of empty slack columns visible. These identify rayfree constraints, which were found to span 10 dimensions. Out of the total of 54 polytope dimensions given by the number of variables, that leaves a ray space of 44 dimensions. As there were 44 linearly independent rays found by the peripheral ray search, this confirms that the ray matrix is vectorially complete. The dominant slack entries appearing in a roughly diagonal line, reflect the systematic probing of slack columns from left to right. This probing produced 40 rays, leaving the remaining four rays to be determined as aligned rays, and they are distinguishable at the bottom of the matrix as four rows that do not participate in the diagonal progression.

4.6 Capping Rays and Closure Testing

In some contexts, it is useful to pursue the goal opposite to peripheral rays, namely to find rays that are as 'interior' as possible. One such context was already encountered when discussing avoidance of corner trapping in centre refinement, Section 3.2.4. Another is to construct capping hyperplanes, as illustrated in Figure 1.4 and the goal is to truncate multiple ray directions in an even-handed way.

A mathematical formulation of what 'interior' means is to take it as a ray that has maximal total overlap with all constraint vectors. So, its construction again involves using the constraint set of Eq. (4.3) and an objective that sums all the slack values, but this time use LP to *maximise* the objective.

Simple maximisation would fail because if a ray direction exists, the calculated vector x can be increased without limit, its overlaps accordingly increase and so the LP solver will exit denoting the problem as unbounded.

This is easily remedied by restricting all x-components to a range $(-1, 1)$. This places no restriction on the directions explored, and the actual normalisation of the (s, x) vector is in any case enforced retrospectively.

At the other extreme, the trivial zero vector also satisfies the ray condition but need not be excluded as it will not normally appear as a maximised solution. The non-negative range constraint on slack variables ensures that if any non-trivial ray exists, it produces a slack sum larger than zero and will be found instead.

Instead, this now serves a useful purpose: if an LP calculation that maximises the slack sum produces the zero vector, it proves that no non-trivial ray solution exists. In other words, the polytope defined by constraint matrix C is closed in this case.

This fact can be exploited to produce a complete set of capping rays. Having calculated the first capping ray, its ray vector is appended to C as an additional row. This in effect introduces a capping hyperplane, although no capping radius is required since the ray condition of Eq. (4.1) is independent of the values vector V involved in the full polytope specification. The capping ray calculation is repeated for the expanded constraint matrix and further capping hyperplanes added in this fashion until a trivial zero ray solution appears. At this point, the originally open polytope has become closed, and the set of capping rays is considered to be complete.

As it is easy to see in 2D and 3D examples, a single capping line or plane can truncate multiple ray directions. In higher dimensions, one capping hyperplane is often not enough to cap all rays, but the loop described before usually terminates after only a few iterations. Actual

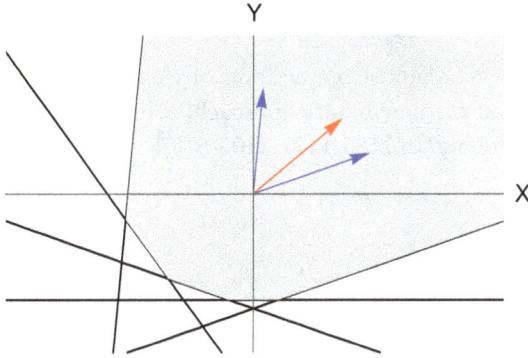

Figure 4.4 Comparison of the calculated peripheral rays (blue arrows) and capping ray (red arrow) for a 2D open polytope specified by constraint lines (black) as the shaded polygon.

examples with ray dimensions up to 100 dimensions, surprisingly only need fewer than four capping rays to attain closure.

As a final graphical example, Figure 4.4 illustrates the distinction between peripheral and capping rays for a partially unbounded 2D polytope. A capping line perpendicular to either blue arrow would truncate the ray directions, but quite unevenly. That distorts the shape of the capped polygon as dictated by the finite boundaries defined by the 'real' constraints, whereas a capping line perpendicular to the red arrow treats both ray directions similarly, and gives a more compact capped polygon without such distortions and a lower aspect ratio.

Another way to avoid the bias that results if a single peripheral ray is used for capping, is to use the entire collection of peripheral rays to introduce multiple capping hyperplanes. Using both blue arrows in the figure for capping, there are two capping lines that intersect and thus add a vertex to the capped polygon. Apart from this, the shape of the capped polygon is quite similar whether using the bespoke capping ray or the pair of peripheral rays.

Using multiple capping hyperplanes of course makes the specification of the capped polytope slightly more complex by adding to the number of constraints although not increasing the dimensionality. In the final analysis, this nevertheless appears more successful in avoiding high aspect ratios. So, direct peripheral ray capping has been adopted as the best approach for high-dimensional SSK calculations to be described in Chapter 7.

Reference

1. J. Gagneur, & S. Klamt, Computation of elementary modes: a unifying framework and the new binary approach. *BMC Bioinformatics*, **5** (2004) 175. https://doi.org/10.1186/1471-2105-5-175.

Chapter 5

Classifying Polytope Facets

5.1 The Facet Hierarchy

In Chapter 1, Section 4, the concept of a polytope facet was briefly introduced. It was shown that in order to reduce the solution space to a compact kernel, it is necessary to distinguish between bounded and unbounded facets so that the capping hyperplanes can be set at radii that prevent the truncation of any bounded facets. This concept is graphically illustrated in Figure 1.4. The overall task addressed in this chapter is the determination of bounded facets in high-dimensional, unbounded polytopes.

Intuitively, boundary facets are just the sections of the constraint hyperplanes that form part of the polytope. Such a facet is unbounded if there is a ray aligned with it, and the work of Chapter 4 on computing peripheral rays supplies the necessary tools to determine that. For a three-dimensional (3D) case such as shown in Figure 1.4, it would be straightforward to run through all constraint hyperplanes and identify the ones without aligned rays as the bounded cases. Even for a solution space (SS) polytope in say 100 dimensions, it would typically have around 150 constraints and it is still quite realistic to test them all since only a single LP is required for each to test for the existence of a ray. However, invariably the outcome is that *all* constraint facets are unbounded! Again, high-dimensional polytopes behave rather differently than in lower dimensional 2D or 3D cases.

To understand that, the first step is to recognise that even in 3D, there are further facets beyond just the boundary planes, namely the edges of the polyhedron. Generally, a facet is defined by two attributes:

- It is located in the intersection of one or more constraint hyperplanes, so the **intersection list** f specifying which hyperplanes intersect, is an identifying attribute.

- The facet is the section of this intersection hyperplane that falls within the polytope, that is, it consists of all points in the intersection that satisfy the feasibility constraints.

More loosely, the term *facet* is sometimes applied to the entire hyperplane specified by an intersection list, in which case the section that satisfies both requirements is called a *feasible facet*.

In N dimensions, the facet definition gives rise to a hierarchy of facets of different dimensionality. Where k hyperplanes intersect, k degrees of freedom become fixed and so a facet associated with an intersection list f of length k has $(N - k)$ dimensions. The value of k is designated as the **facet level**. In particular, the constraint hyperplane facets are level 1 facets and have $(N - 1)$ dimensions, and there are higher level facets up to $k = N$ where all degrees of freedom become fixed and the facet reduces to a single point. So, the polytope vertices are level N facets, and the edges that connect them are level $(N - 1)$ facets. At the other extreme, the polytope itself can be considered as the level 0 facet.

Even in 3D, it is possible that all level 1 facets are unbounded, but there are bounded level 2 facets. Figure 4.3 illustrates such a case, in which all four boundary planes are unbounded, and there are four unbounded edges, but also one bounded edge.

So, it is clear that the search for bounded facets will have to extend over all facet levels. Even though the number of level 1 facets are generally quite small even for high N values, that is not true for higher levels as demonstrated by the previously discussed combinatorial explosion in the number of level N facets (vertices).

For M boundary hyperplanes, the number of hyperplane intersections at a given level k is easily calculated as the combinatorial coefficient $^{M}C_k$. But as even the 2D example in Figure 4.4 illustrates, not all intersections are in the feasible region, and not all of those that are, are bounded. The number of feasible bounded facets (FBFs) obviously depends on the detail of each polytope, but the general trend of how the FBF count depends on the level, can be established by the following argument.

Given the set of constraint equations $C \cdot x \leq V$, the points that belong to boundary facet j satisfy all these inequalities, but strict equality for row j in matrix C. Similarly, the feasibility condition for a higher level facet specified by an intersection list is obtained by changing all rows specified in the list to strict equalities. Every such change makes the constraint set increasingly more discriminating, so the fraction of distinct intersection lists that satisfy the feasibility requirement *decreases* monotonically with the facet level k. At level 0, there is only one facet (the polytope itself) and it is by definition feasible, so the fraction value is 1. At level N, there is a large number of intersections but only a small fraction of them will satisfy the more stringent feasibility requirement containing N strict equalities.

The same argument can be applied to the constraints $C \cdot x \leq 0$, that define rays. For a ray to be aligned with a facet, it has to be orthogonal to all the constraint vectors in the intersection list associated with the facet. So again the corresponding rows in the ray constraint set become strict equalities, and it follows that as the level increases, the fraction of intersections that allow rays aligned to the facet decreases monotonically. Since a bounded facet is one that contains no rays, the fraction of bounded facets *increases* monotonically with the facet level. When considering an unbounded polytope, at level 0, there are no bounded facets so the fraction is 0, whereas at level N, every intersection is a geometric point and is by definition bounded so the bounded fraction is 1.

These relationships are schematically represented in Figure 5.1, with the sigmoidal curve shapes of part (a) based on trials with low-dimensional polytope examples. The fraction of facets that are both feasible and bounded is represented by the small blue shaded overlap area in Figure 5.1(a). It is seen that FBF facets can only be expected to occur near the middle of the level range, and even then, only represent a small fraction of intersections – how small, is an interesting question.

Figure 5.1(b) plots the combinatorial intersection count for the 138 constraints in 109 dimensions that define the Reduced Solution Space

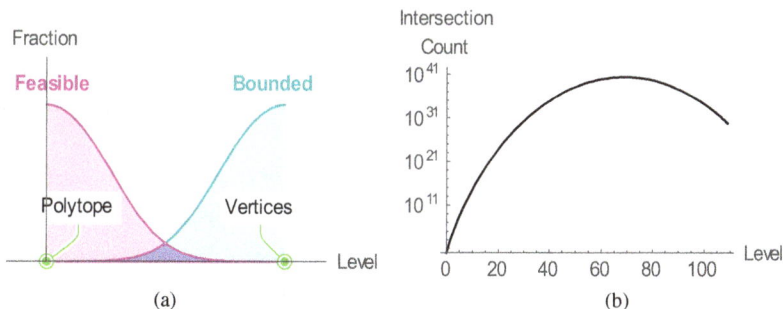

Figure 5.1 Estimating (a) the fraction of facets that are feasible and bounded and (b) the total number of hyperplane intersections for the Geobacter RSS.

(RSS) of the previously discussed Geobacter model, a typical medium-sized metabolic model. Combining the two graphs leads to a startling conclusion. Even if the FBF fraction is as small as one in a million, the total number of them will still be a computationally overwhelming number around 10^{35}. On the other hand, if the FBF count is a manageable number, the fraction of intersections that yield an FBF has to be negligible. These scenarios present severe challenges to the prospect of practically finding a representative sample of FBF's in the first case, or even finding any FBF at all in the second.

As it turns out, at least in the Geobacter case, it is the second scenario that applies. Using the methods presented below, it is found that this model happens to have just 960 FBFs of interest, a negligible fraction of all facets. In other models, even smaller numbers are often found.

Unsurprisingly, quite an elaborate conceptual framework needs to be established first in order to identify sufficient FBFs; that is the aim of the following sections.

5.2 Relationships and Facet Trees

The explicit feasibility constraints for an individual facet at any level as formulated previously, imply that each facet of a polytope is again a convex polytope, but in fewer dimensions.

They are also related; for example, in a 3D polyhedron, each planar first-level boundary facet is bounded by edges, that is, second-level facets, and these in turn by vertices, that is, level 3 facets.

This can be generalised by considering the intersection list $f = \{m_1, m_2, \ldots m_k\}$ of a level k feasible facet, where the m_i are sequence numbers that identify constraints, that is, rows in C. Any subset f' of $(k - 1)$ entries in this list defines a lower level facet. The collection of level k facets that share f' as a subset, define *sub-facets* of the level $(k - 1)$ facet that f' identifies. Put in another way, starting from a level k facet, all its sub-facets inherit its intersection list but adds a distinct additional constraint hyperplane to the list.

The inheritance suggests that facets can be organised into a family tree structure, where facets are nodes and each node branches into sub-facets at the next higher level of the tree. The complication is that each sub-facet has multiple parents at the previous tree level, just as in the polyhedron, any edge is shared by the two planes that intersect to form the edge. This gives rise to a cross-linked structure that is not really a tree structure at all.

A much simpler representation that avoids this is created instead by associating each node with the number of the additional constraint that is added to reach the next facet level. This gives rise to a proper tree structure.

In this tree, a facet is represented by the constraint number list that is picked up when traversing the path from the root to a particular node along the tree. The simplification of the tree structure achieved by associating a facet with a path rather than a node, is because the same path is reused many times for different facets.

Although this tree is called a facet tree, each path defines a hyperplane intersection rather than a facet. Each path terminates at a unique node, and in that sense not only the path, but also each node is associated with a different intersection list. Only a subset of nodes represents feasible facets though, because not all intersections satisfy the feasibility test.

The facet tree is not intended to give a detailed characterisation of the polytope geometry. Its purpose is best considered as merely a framework that allows systematic, level-by-level exploration of all its facets.

5.2.1 *The Structure of the Facet Tree*

A small facet tree example is shown in Figure 5.2. The root node is labelled with the empty list {} and represents the polytope itself, at

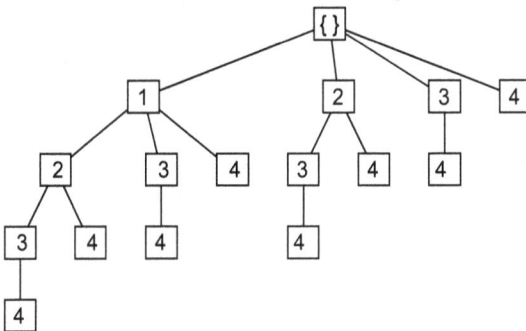

Figure 5.2 Facet tree example, for a four-sided open cone in four dimensions. Following computational convention, the tree is presented upside down, that is, the root (level 0) is at the top and the highest tree level at the bottom.

level 0. For each node in the tree, the rule is that the numbers allocated to its children in the next level are only those numerically following on its own number. Each tree level corresponds with a facet level. As customary, the tree is presented 'upside down', that is, higher tree levels are lower on the page.

The stated node numbering rule ensures that along any path, all the numbers that are picked up maintain their numerical order. This in turn ensures that intersections are not duplicated and there is a one-to-one correspondence between nodes and intersection lists. The correspondence with facets is more complicated and is discussed below.

Inspection of Figure 5.2 shows an interesting structural feature. Each subtree formed by each of the main branches (those branching off the root node) gets repeated at the next tree level on each main branch to its left. The same is true of subtrees at higher levels as well. This makes for a highly asymmetric tree structure where subtrees terminate on increasingly higher tree levels as branches are considered from right to left. The leftmost edge of the tree defines the only path that picks up all M constraint hyperplanes, and so determines the maximal tree depth as M.

The intersection lists associated with all nodes at level k, cover all subsets of length k that can be chosen out of the sequence number list $\{1, 2, ..., M\}$ and so there are a total of $^M C_k$ nodes at level k of the tree. In Figure 5.2, for example, the number of nodes at increasing tree levels is 1, 4, 6, 4 and 1, which is recognisably the combinatorial sequence $^4 C_k$. The maximal number of nodes per level occurs at the middle level $k = \frac{1}{2} M$. This defines the total width of the tree as $^M C_{M/2}$.

The total number of tree nodes is given by adding the level counts and that is easily done using the binomial theorem:

$$(1+x)^M = \sum_k {}^M C_k \, x^k \quad \Rightarrow \quad \sum_k {}^M C_k = 2^M \tag{5.1}$$

Full traversal of the facet tree guarantees that each constraint hyperplane intersection will be encountered exactly once. But the doubling of tree size with each increment of the constraint count means that this becomes impractical for quite modest polytopes, with say $M = 20$ sides.

Much of the following work is aimed at restricting the extent of the traversal needed to find the FBF's of a polytope. An obvious starting point for this is to note that the N-fold intersection of boundary hyperplanes in N dimensions defines a geometric point, such as a polytope vertex. So, it should be sufficient to terminate the traversal at level N of the tree in the usual case that $N < M$. Generally, all higher level nodes will be infeasible, except in cases where polytope symmetry causes multiple vertices to coincide. Even then, each of the coinciding vertices will already have been separately encountered at level N, so termination at that level will not miss any facets.

An important feature of the facet tree structure is how it reflects ancestry relationships between facets. Assuming the 4D cone to be symmetric, each of the four 'sides' (level 1 facets) has the same arrangement of sub-facets, but in Figure 5.2, the node for facet $f_1 = \{4\}$ at the right of the tree shows no descendant whereas for facet $f_2 = \{1\}$ at the left, all its descendants $f_2 = \{1,2\}$, $\{1,3\}$ and $\{1,4\}$ are explicitly shown. The corresponding descendants of f_1 are nevertheless all present, at the nodes or paths represented by $\{3,4\}$, $\{2,4\}$ and $\{1,4\}$ respectively. Note that all of these are located to the left of their common parent node $\{4\}$.

The general rule that results from the asymmetric tree structure is that all descendants of a facet not explicitly shown above its node (in terms of tree levels, not placement on the page!) as descendants on the tree, are located above it and to its left, on subtrees rooted at the same tree level as the chosen node.

The reason why this is true in general is demonstrated by taking an explicit example using Figure 5.2. Consider the node of the facet specified by its intersection list $f = \{1, 3\}$. All its sub-facets have f as a subset of their own lists. However, no path to the right of this node can pick up the entry 1 on this list, and so all non-explicit sub-facets (in this case the single node $\{1, 2, 3\}$) lie on its upper left. Similarly, at higher levels and on larger trees, all entries in the intersection list up to the penultimate one, rule out all nodes to the right as sub-facets of a given node.

A result of this observation is that only for nodes on the left-hand boundary of the tree, is it in fact true that all sub-facets are explicit descendants because there are no subtrees further to the left.

5.2.2 *Polytopes and Facet Trees Have a Many-to-Many Association*

There are two parts to this assertion. Firstly, the same facet tree can represent different polytopes, distinguished, for example, by different subsets of nodes being feasible, as long as they have the same number M of constraint hyperplanes.

For example, Figure 5.2 equally well represents a 2D square, trapezoid or quadrilateral, a four-sided 3D cone or a 4D cone.

For the 2D square, levels zero and one are feasible, only four out of the six nodes in level 2, and none in levels 3 and 4. A trapezoid adds one more feasible node in level 2 to this list, and for a general quadrilateral, all six nodes become feasible but levels 3 and 4 remain infeasible.

For a 3D four-sided cone with a square cross-section, all six level 2 nodes are feasible; while opposing sides of the 2D square do not intersect, the corresponding sides of the 3D cone do intersect at the apex of the cone. In fact, since all four sides intersect at the apex, it is a level 4 facet represented by the single tree node at that level. Moreover, all four level 3 nodes also represent coinciding feasible facets, namely the apex point at which each subset of three sides of the cone, also intersect.

In this example, there are seven distinct tree nodes that all represent the same geometrical facet, including one beyond the polytope dimension $N = 3$. It also demonstrates that nodes at the same tree level can represent facets with different dimensionality: four of the level 2 facets represent edges of the 3D cone, while the other two represent a vertex.

Clearly, the simple one-to-one correspondence between nodes and intersection lists does not transfer to facets, but it remains true that all facets are encountered in a full tree traversal (only possibly multiple times, in cases where they happen to coincide geometrically).

Applied to a 4D cone, all nodes in the facet tree of Figure 5.2 can represent distinct feasible facets.

The second part of the assertion in the heading is the reverse statement: a given polytope can be represented by multiple facet trees, and this is also true.

To see that, consider that in the intersection list for any given facet, the entries can in principle be listed in arbitrary order without changing the facet it describes. In fact, the ordering of constraint equations is itself arbitrary; reordering them in any way still represents the same polytope.

Suppose that a fixed association is created between the initial row numbers in the constraint matrix C and the corresponding constraint hyperplanes. Let the rows of C be rearranged (e.g., to the ordering [4 2 3 1] in the 4D cone example) and this new ordering be used to build a new facet tree.

This tree will give an entirely equivalent representation of the polytope facets, but the node representing an individual facet will occupy a different position in each of the two trees and show different relationships to other nodes.

For the 4D example as shown in Figure 5.2, the level 1 facet $f_1 = \{4\}$ at the right of the tree shows no descendant whereas for facet $f_2 = \{1\}$ at the left all its descendants $f_2' = \{1,2\}$, $\{1,3\}$ and $\{1,4\}$ are shown. With the new ordering, facet f_2 will now appear at the right with no descendant shown while f_1 appears on the left showing the descendants $f_1' = \{4,2\}$, $\{4,3\}$ and $\{4,1\}$.

Generalising, for a polytope with M boundary hyperplanes there are a total of $M!$ distinct facet trees, each associated with a distinct ordering of the constraint matrix rows. All these trees are completely equivalent in that each tree represents all intersections with a one-to-one correspondence to its nodes; but each having a unique arrangement of where a particular facet is positioned on the tree. Each tree displays a different subset of the ancestry relationships explicitly, and the complete set of such relationships is represented over the collection of equivalent trees rather than any single tree.

One way in which the diversity of trees can be exploited, is that for a given level k facet f of interest, a tree can be selected such that all its sub-facets appear in the subtree rooted at node f. To achieve this, the chosen facet only needs to appear on the left-hand boundary of the tree. The sequence of entries along this leftmost path is by construction identical to the sequence that defines a particular tree. So as long as the length k intersection list f forms the initial k entries on the sequence of the ordering that defines the tree, it will appear on the left-hand boundary. Multiple equivalent trees satisfy that requirement, and in all of those sub-facets of f appear explicitly in the subtree rooted at node f.

5.3 Inheritance of Boundedness and Feasibility

For a simple example like the tree in Figure 5.2, it is straightforward to traverse the tree, and at each node test for feasibility and boundedness in order to find all FBF's that belong to a particular one of the various polytopes modelled by the tree. However, even restricting traversal to levels up to the spatial dimension N, the exponential explosion of tree size as determined by M, makes this impractical for all but very small polytopes.

Such a search might be streamlined for larger polytopes if the need for tests can be avoided by knowledge of how these traits are inherited by sub-facets. For example, since all points on a sub-facet also belong to the parent facet, it is clear that a bounded facet can only have bounded sub-facets, and so boundedness testing can be skipped for them.

It is tempting to similarly conclude that a feasible facet can only have feasible sub-facets, but that is misleading since as a counterexample, the polytope itself is feasible but not all its boundary intersections are.

The key difference is that in the case of feasibility testing, a positive outcome only asserts that there exist some points in the hyperplane that are feasible, not that all points are. It is entirely possible that two hyperplanes that each tests positive for containing a feasible facet, may only intersect outside of their feasible subsets. In this case, feasibility is not inherited by the subfacet.

On the other hand, if the feasibility test fails, that means that the candidate facet hyperplane does not contain any feasible points, so its intersection with any other hyperplane will also be infeasible.

The case that the boundedness test fails is a counterpart to this. As already shown for example in Figure 4.3, a polytope can be unbounded but still possess bounded sub-facets.

These considerations establish the following inheritance rules, first for descendants and then for ancestors:

- The sub-facets of an infeasible 'facet' are also infeasible
- The sub-facets of a feasible facet do not inherit its status
- The sub-facets of a bounded facet are also bounded
- The sub-facets of an unbounded facet do not inherit its status
- All parents of a feasible facet are also feasible
- Parents of an infeasible 'facet' may be either feasible or infeasible
- All parents of an unbounded facet are also unbounded
- Parents of a bounded facet may be bounded or unbounded

As an application of the rules, consider an FBF search strategy that traverses the facet tree starting from the root node (the polytope), for which the status is that it is feasible but unbounded. The rules above imply that node feasibility testing along a path may be stopped at any node that tests as infeasible, as all further nodes along this path will be infeasible. Similarly, boundedness testing may be stopped at any node that tests as bounded, with all subsequent nodes along this path classified as bounded.

5.4 FBF Base Level Facets

It is also worth investigating the situation if an FBF has been detected at a node f, perhaps by another method than systematic tree traversal.

Since it is feasible, all its ancestors right back to the root are also feasible. But not all of them will be bounded. So, it is possible to trace the ancestry of f moving up along the unique path that leads to f, and test each node for boundedness. Sooner or later, a node will be encountered that is unbounded (at worst, the root node). All further ancestor nodes up to the root will also be unbounded.

So, the last bounded node that was found in this way, is also the first and lowest level FBF encountered along the path starting from the root. It is important enough to be specially designated as the **Base Level Feasible Bounded Facet** or **BFBF**. Not only the original node f, but all other descendants of the BFBF that are feasible and bounded, are also FBFs.

One reason that a BFBF is important, is understood by recalling that the purpose of finding FBFs is to find tangent capping hyperplanes for the unbounded polytope, defined such that they do not intersect any FBF at more than a single tangent point. If a capping hyperplane satisfies this requirement for the BFBF, it automatically also satisfies it for all its FBF descendants.

So, the stated purpose of finding all FBFs of the polytope can now be reformulated as *just finding all BFBFs*. That is obviously a huge simplification of the task. In terms of Figure 5.1, rather than having to find all FBFs at all levels covered by the blue triangular region, only those at lower levels may suffice.

The process of moving down the tree levels from the FBF node F to its BFBF ancestor, is described as **backtracking** in subsequent work. Obviously many FBF's can share the same BFBF ancestor, but there may be multiple BFBFs on a specific tree. So FBFs can be classified in different families, each with its own unique BFBF ancestor. The collected BFBFs do not necessarily occur all at the same facet level, but a relatively small set of BFBF levels at the lower end of the level range is usually found to account for all known FBFs.

5.5 The Progenitor Facet

Once all BFBFs 's have been found it is of interest to pursue the ancestry of this collection of facets too. For any two BFBFs, moving down the tree levels will sooner or later yield a common ancestor. As each BFBF is by definition at the lowest level that yields a bounded facet along its path, this common ancestor is sure to be an unbounded facet.

The unbounded facet that forms the highest level common ancestor of the entire collection of BFBFs, is denoted as the **progenitor facet**.

The progenitor is guaranteed to exist, since the polytope itself (i.e., the root node) is the ancestor of all facets and so is the progenitor of last resort. However, in the high-dimensional polytopes encountered in metabolic models, the progenitor facet can occur at a relatively high level, although obviously at a lower level than any BFBF.

So, the progenitor is an *unbounded* but *feasible* facet that has all BFBF's and consequently all FBF's as descendants, but not all its descendants are either feasible or bounded.

The process of finding the progenitor also involves moving down the facet tree from known facet nodes (the BFBFs), and is called **traceback** to distinguish it from the backtracking mentioned previously.

The first reason that the progenitor facet is important for the SS kernel calculation is that it allows reduction of the facet tree size. Section 5.2 showed that it is possible to ensure that all sub-facets of the progenitor, which by definition means all FBFs, are placed on the subtree rooted at the progenitor node. Any facet tree that has the intersection list belonging to the progenitor as its initial entries in the hyperplane ordering that defines the tree, will place the progenitor node on its left-hand boundary and so have all sub-facets on its subtree.

So once the progenitor is known, the empty listed, root node as in Figure 5.2 can be replaced by a progenitor list node, and only subsequent entries in the tree ordering list used to generate further nodes for the tree. For a progenitor at level p, the tree size is reduced from 2^M to 2^{M-p} nodes. Values of the order $p = 50$ is not uncommon, so the simplification can be significant.

In most cases, there is a second reason that the progenitor is useful, namely that it allows a reduction of the spatial dimensions by coincidence capping. Suppose that a ray exists that is orthogonal to the progenitor facet. Then a capping hyperplane can be constructed that is orthogonal to that ray, and is located at a distance R (the capping radius) from the coordinate origin O, which is chosen inside the progenitor hyperplane. Now consider any point X in the progenitor hyperplane and another point Y located in the capping hyperplane. Then the vector XY that connects these points, will generally contain a component along the ray vector, because if it does not it would mean that XY lies entirely in the progenitor plane. This would only be true if the progenitor and capping hyperplanes coincide.

Since all FBFs are sub-facets of the progenitor, it follows that R may be reduced to zero without the capping hyperplane intersecting any FBF. In this case, it becomes a coincidence capping plane. As already described in Chapter 4, when discussing prismatic rays as in Figure 4.1, this amounts to a projective transform that reduces the spatial dimensions and eliminates any constraints that become redundant.

The most efficient way to combine the two uses of the progenitor, has been found to apply coincidence capping if an orthogonal ray exists, otherwise just the reduction to the progenitor rooted subtree. If coincidence capping was done, the ray matrix calculation and further search for BFBFs is performed in the resulting lower dimensional space.

To put these reductions into perspective, for the Geobacter example with an RSS of 138 constraints in 109 dimensions, BFBFs are found at levels 76, 77, 79 and 80 and their progenitor is found at level 53. Just readjusting the root reduces the tree size by a factor of $2^{53} = 9 \times 10^{15}$, although the search is still conducted in a 109-dimensional flux space. But in fact, an orthogonal ray does exist in this case, and coincidence capping reduces the polytope to just 85 constraints in 56 dimensions with a correspondingly even smaller tree size as well.

The previous discussion skipped some crucial features: how to calculate the progenitor facet, and how to establish if there is a ray that is orthogonal to it. The first of these will have to wait for a later chapter, but the second point can be addressed now as it is just an extension of the procedure to find a capping ray that was presented in Chapter 4, Section 6.

To recap, that entailed transforming the inequality $C \cdot x \le 0$ that defines rays, to a set of equations by introducing a non-negative slack variable s_j for each of the M rows in the constraint matrix C. The sum of slack variables is taken as the LP objective and is maximised, subject to the N components of the flux space vector x being restricted to the range $(-1 \le x_i \le 1)$.

In addition to that, further constraints are now added to keep the ray orthogonal to the progenitor facet. Let the progenitor facet be defined by its intersection list $p = \{m_1, m_2, \ldots, m_k\}$. The rows of the constraint matrix are partitioned into two sets: P and NP where P includes all rows that belong to the set p, and NP are the rest, that is, $C = \text{Join}[P, NP]$. The vector basis for the intersection of the listed constraint hyperplanes, is given by $B = \text{NullSpace}[P]$. So, the combined constraint equations are given by

$$\begin{bmatrix} I & C \\ 0 & B \end{bmatrix} \cdot \begin{pmatrix} s \\ x \end{pmatrix} = 0 \tag{5.2}$$

and LP maximisation of the sum of s-components is performed as before. The actual solution is not required for the coincidence capping; just the existence of a non-trivial solution establishes that the orthogonal ray exists and therefore projection to the progenitor hyperplane can be performed.

In practical calculations, the ray matrix calculation precedes finding the progenitor. A single matrix multiplication establishes if any of the rays contained in the ray matrix are orthogonal to B. Only if that test fails, is it necessary to solve Eq. (5.2) to definitively determine if a ray that is orthogonal to the progenitor exists.

5.6 Feasibility and Boundedness Testing

Leaving aside the details of the tree traversal, in order to find BFBF's it is necessary at each tree node that is visited, to test if the associated hyperplane intersection is both feasible and bounded. The large number of nodes involved makes it imperative that this testing should be as efficient as possible.

As done for the progenitor, it is convenient when testing a facet specified by the intersection list f to partition the constraint matrix row-wise into the sub-matrix F of rows that belong to f (the facet rows), and the complementary non-facet rows NF. These are referred to as facet and non-facet constraints respectively. The values vector is similarly partitioned into V_f and V_{nf}.

The facet is feasible if there exists a point x_f such that:

$$F \cdot x_f = V_f \quad ; \quad NF \cdot x_f \leq V_{nf} \tag{5.3}$$

The first requirement ensures that x_f is located in the intersection and the second that it is feasible.

Boundedness is established by testing if a ray $\hat{r}$ exists that is aligned with the facet, that is, is orthogonal to all the facet constraint vectors, listed in f. So $\hat{r}$ is a solution to

$$F \cdot \hat{r} = 0 \quad ; \quad NF \cdot \hat{r} \leq 0 \tag{5.4}$$

In this case, the facet is unbounded if the solution exists and bounded if not.

In each of Eqs (5.3) and (5.4), the parts can be combined and solved as a single LP once a suitable objective vector is introduced. But solving two LPs at each tree node requires considerable computational effort and severely restricts the tree size that can be practically traversed, so approaches that are more elaborate but also more efficient are proposed below.

5.6.1 *Feasibility*

The first part of Eq. (5.3) is just an algebraic equation and can be solved as in Eq. (1.5) by using the Moore–Penrose pseudoinverse F^+:

$$x_f = F^+ \cdot V_f + \left(1 - F^+ \cdot F\right) \cdot W \tag{5.5}$$

$$x_f = \Phi + Y_f \tag{5.6}$$

According to the underlying theory of the pseudoinverse, $\Phi = F^+ \cdot V_f$ is the minimum norm solution. This can be geometrically interpreted as the point in the intersection that is closest to the coordinate origin.

In Eq. (5.5), W represents an arbitrary vector, and $(1 - F^+ \cdot F)$ is a matrix that projects any vector on to the null space of F, which in this case is the hyperplane intersection itself. So, Y_f is simply an arbitrary vector that belongs to the intersection hyperplane.

However, this solution only exists if the following further condition is satisfied:

$$F \cdot \Phi = V_f \tag{5.7}$$

If Eq. (5.7) does not hold, there is no solution for x_f. This accounts for the case that the hyperplanes specified by the intersection list f do not geometrically intersect. Note that calculation of F^+ and hence Φ only involves standard matrix algebra, so if they fail to satisfy Eq. (5.7), facet f can be declared immediately as infeasible and all LP testing is avoided. A large fraction of tree nodes falls in this class and so benefits from taking the pseudoinverse approach.

If Eq. (5.7) holds true, feasibility requires that the second part of Eq. (5.3) can be satisfied too by a suitable choice of Y_f. Substituting Eq. (5.5) into Eq. (5.3) leads to the pair of requirements:

$$F \cdot Y_f = 0 \quad ; \quad NF \cdot Y_f \le V_{nf} - NF \cdot \Phi \tag{5.8}$$

This pair of relations could be unified into a single constraint set and solved by LP, but explicit use of the first equation is avoided and the matrix size reduced by substituting Y_f in favour of W as used in Eq. (5.6). Then, Eq. (5.8) reduces to:

$$NF \cdot \left(1 - F^+ \cdot F\right) \cdot W \leq V_{nf} - NF \cdot \Phi \qquad (5.9)$$

$$G \cdot W \leq S^\Phi \qquad (5.10)$$

The second equation merely introduces new symbols, defining an effective constraint matrix G which has a row count equal to the number of non-facet constraints in NF. The vector S^Φ renames the right-hand side vector of Eq. (5.9), and the notation chosen will be clarified in the following.

So, in cases where Eq. (5.7) is satisfied, an LP solution subject to the constraints contained in G is attempted. The objective to be optimised is immaterial since only the existence of a solution is of interest; in practice the objective vector is chosen as the zero vector. If a solution exists, the node intersection list gives a properly feasible facet, otherwise it is infeasible and not strictly a facet.

5.6.2 Boundedness

Once again, the goal is to avoid LP testing where possible and for boundedness testing the ray matrix is the key tool in this regard. Recall that the first M columns of this matrix lists the (negative of) overlaps of the ray vector in each row, with each of the constraint vectors. For a tree node with the associated intersection list $\{m_1, m_2, \ldots m_k\}$, inspecting this subset of the ray matrix columns gives information on how the known rays relate to the supposed facet.

A ray that is aligned with the facet has a zero overlap with each of the constraint vectors in the intersection list. So, if the submatrix of ray matrix columns in the intersection list contains any zero rows, an aligned ray is known and the facet is definitely unbounded.

When it has no zero rows, the facet is *likely* to be bounded. Note that as all entries in the submatrix are by construction non-negative, no convex combination of non-zero rows can form a zero row, and so aligned rays cannot 'hide' in a sub-matrix without zero rows.

If the ray matrix contained a convex complete ray basis, as it would if it listed all extreme rays, absence of a zero row in the sub-matrix would conclusively prove that the facet is bounded. But with a ray matrix constructed as described in Chapter 4 that is merely vectorially complete, the absence of a zero row only ensures that there is no aligned ray formed from a convex superposition of the known rays.

So, to definitely establish boundedness in this case it is still necessary to perform a ray finding calculation that in addition to the ray requirement $C \cdot x \leq 0$ sets all slack variables that belong to the facet columns $\{m_1, m_2, ... m_k\}$ to zero. This is easily incorporated into the peripheral ray LP calculation as described in Section 5 of Chapter 4, by excluding the facet columns from the probe column list it uses. The reduction in the number of probed columns, means that ray detection becomes quicker in high-level facets.

If this calculation succeeds in finding a new ray aligned to the facet the tree node represents, this ray is added to the ray matrix and used in all subsequent facet tests. In practice, it is found that this only happens very rarely, but is a reason why the final ray matrix after the tree traversal sometimes turns out to be vectorially overcomplete.

So, in boundedness testing, unbounded facets are mostly detected merely by inspecting the ray matrix, and LP testing is only required to definitely confirm the status of the subset of facets that already passed the first hurdle presented by the ray matrix.

To make the inspection test as discriminating as possible, the ray matrix should both include as many rays as possible, and the rays that are included should be as peripheral as possible to make the ray matrix sparse and facilitate the presence of zero rows in submatrices. Both of these requirements were guiding principles in the ray matrix construction set out in Chapter 4.

5.6.3 *Integrating Feasibility and Boundedness Testing*

Further gains in efficiency can be made by recognising that testing the two qualifying aspects of an FBF are not really independent. Obviously, if the first aspect tests negative, there is no need to perform the second test. This raises questions about whether feasibility or boundedness should be tested first. It seems plausible to apply either the quickest test or the one that is most likely to fail first. But both of these aspects may vary with the facet level and is hard to predict in advance for a specific facet.

However, there are also links between the tests that arise from the algebraic similarity of their defining Eqs (5.3) and (5.4), and exploiting this the tests can be integrated, largely avoiding the sequence problem.

Since feasibility testing can cheaply eliminate non-intersecting hyperplanes, that is a good place to start. So given a facet list $f = \{m_1, m_2, \dots m_k\}$, Eq. (5.7) is first tested and if it fails it is classified as {feasibility, boundedness} = {infeasible, unresolved}.

Otherwise, testing moves on to Eq. (5.9) and the vector S^Φ is evaluated. By definition this requires calculating the vector:

$$S^\Phi = V_{nf} - NF \cdot \Phi \qquad (5.11)$$

Consider the special point $x_f = \Phi$. By virtue of Eq. (5.7) it already satisfies the first feasibility condition in Eq. (5.3) while the second reduces to the inequalities $NF \cdot \Phi \leq V_{nf}$. If these are converted to equalities by introducing slack variables, the resulting vector of slack variables is given by Eq. (5.11). This explains the choice of notation S^Φ, indicating the slack variable column vector associated with point Φ. In particular, for this point to be feasible, all elements of S^Φ need to be non-negative.

Starting from a given f, the vector S^Φ as calculated from Eq. (5.11) using the pseudoinverse F^+, may or may not have negative elements. If it does not, it means that $Y_f = 0$ satisfies Eq. (5.8) so that according to Eq. (5.6) $x_f = \Phi$ is a feasible point and so f can be classified as {feasible, unresolved}.

More generally, S^Φ will contain some negative entries and Φ by itself is not feasible, but there may still exist Y_f vectors that give feasible solutions. The known ray vectors can be invoked to investigate that further, and also to help resolve the boundedness status. To facilitate the discussion, a distinction is made between a facet that is **ray free**, meaning there are no *known* rays aligned with it, and the stricter description that it is **bounded** if no aligned rays exist *at all*. If a facet is not rayfree, it is definitely unbounded, but the opposite is not true.

Comparing the facet ray condition Eq. (5.4) with the feasibility condition Eq. (5.8), it is seen that any positive multiple of such a ray satisfies the first feasibility condition for Y_f. So, candidate Y_f solutions can possibly be constructed from known ray vectors, depending on whether they satisfy the second part of (5.8) as determined by the entries in S^Φ.

A number of different cases can arise:

(1) Case 1: Facet f is ray free by inspection of the ray matrix.
  (a) $S^\Phi \geq 0$, that is, all elements are non-negative, Φ is feasible and f is {feasible, ray free}.
  (b) S^Φ has negative entries. As there are no known rays, the status of f is {unresolved, ray free}.
(2) Case 2: Ray matrix inspection yields known aligned rays and $S^\Phi \geq 0$. In this case, Eq. (5.4) applies and is a stricter condition than Eq. (5.8), so f is {feasible, unbounded}.
(3) Case 3: Ray matrix yields known rays but S^Φ has negative entries.
  (a) S^Φ has only one negative entry, in row j, of value $-\alpha$.
    Suppose that row q in the ray matrix has a non-zero entry with value β in slack column j. This means that $C_j \cdot \hat{r}_q = -\beta$ and so the vector $Y_f = \gamma \hat{r}_q$ will still satisfy Eq. (5.8) provided that we choose $\gamma \geq \alpha/\beta$.
    So, if ray matrix inspection yields any non-zero entries in column j, facet f is {feasible, unbounded} else {unresolved, unbounded}.
  (b) S^Φ has multiple negative entries in the set of rows $J = \{j_1, j_2,...\}$. If there is a single row q in the ray matrix, which has non-zero entries in all columns J, ratios γ as for case (a) can be constructed for each element of J. The maximum of these can then be similarly used to construct a valid solution Y_f as a multiple of the ray. If there is no single suitable row q, but there is a set of rows that between them have non-zero entries in all columns that belong to J, any convex combination of these rows will give a ray with non-zero entries in all columns that belong to J and as before a suitable multiple of this ray vector will give a valid solution Y_f.
    So, the outcome is that facet f is classified as {unresolved, unbounded} if there is at least one column of the ray matrix, belonging to set J, for which all entries are zero, otherwise the outcome is {feasible, unbounded}.

The case classification only requires ray matrix inspection, but may need further resolution by LP calculation. Bearing in mind the inheritance rules, the various scenarios that arise are dealt with as follows:

A. {infeasible, whatever} — No further testing, terminate at this node
B. {feasible, unbounded} — No further testing, proceed to next node

C. {unresolved, unbounded} — Do reduced LP feasibility test Eq. (5.10), then perform A or B according to the outcome

D. {unresolved, ray free} — Do reduced LP feasibility test Eq. (5.10), then perform A or E according to the outcome

E. {feasible, ray free} — Do LP boundedness test as in Section 5.6.2 , then terminate if positive, else proceed to next node.

Note that traversal of any tree branch starts from scenario B, continues as long as this is repeated, and stops only if either A or E is encountered. So, for most nodes LP testing is not needed; some require the reduced LP feasibility test, and with rare exceptions (where the ray matrix is inadequate) the successful location of an FBF requires both LP tests.

This is a vast improvement over direct testing based on Eqs (5.3) and (5.4). Even compared to using the improved tests separately, integrated testing still further reduces computation times by a factor of 2 or more in test problems.

The final result is that traversal of hundreds of thousands, even millions of tree nodes becomes realistically achievable, when implementing the integrated node testing for searching FBFs.

Chapter *6*

The Search for Base Level Feasible Bounded Facets

6.1 Systematic and Random Searches

Chapter 5 indicated that the sheer number of facets of the solution space polytope for a realistic metabolic model, and the comparative rarity of bounded facets, makes it a formidable challenge to find these. The concept of a facet tree gives the framework for a systematic search and various strategies were introduced to reduce the search space and make the testing of a particular candidate facet efficient. Even so the task remains far from easy.

Traversal of tree structures is a recurring problem in computational science. There are multiple standard algorithms for traversing binary trees, in which there are two branches at each node. But the facet tree shown in Figure 5.2 is a multinomial tree, which is an altogether different proposition. Some familiar examples of multinomial trees are encountered in programs that play board games like chess or go, where a choice between multiple possible moves has to be made at each decision point. This is usually addressed by invoking artificial intelligence methods, which learn from previous examples to optimise these choices.

However, for facet finding, that does not seem to be a promising approach. There is no reason to expect that the location of base level feasible bounded facet (BFBF) nodes within the same facet tree for different polytopes of different shapes and even dimensions would have any correspondence that could be picked up from learning examples. Even a single polytope is equivalently represented by a multitude of trees with different BFBF locations, so whatever choice of a particular tree is made for one polytope, the choice made for a different polytope is not likely to bear any resemblance to it.

On the other hand, linear programming (LP) algorithms demonstrate that even faced with an overwhelming range of possible choices, a clever strategy can successfully arrive at an optimal choice. The Simplex algorithm in particular, does in principle navigate its way among astronomical numbers of nodes of the facet tree (albeit only at a single level, the level of polytope vertices) to find one that minimises or maximises a chosen objective. The secret is that the LP objective defines a scoring criterion by which the nodes can judged and traversal decisions made.

Although this is encouraging, the BFBF search needs to cover all facet levels, and in addition needs to find not just a single example, but all nodes that qualify. That is a somewhat more daunting task.

The first step is to establish a scoring system to take the place of the objective used in LP. Then some randomisation procedure is needed to allow multiple paths to be explored so additional BFBFs can be found. Finally, the lessons learned from this semi-random search will be used to give an alternative, systematic tree search algorithm that can be employed on moderately sized reduced solution space (RSS) polytopes.

6.2 A Scoring System for Constraint Hyperplanes

It is instructive to discuss the BFBF search problem in the vernacular of a board game. Consider such a game played on a board that is laid out with a facet tree such as in Figure 5.2. A player starts by putting his marker on the root node of the tree, and at each turn chooses to move to the next tree level along an available branch. All nodes along his path need to be feasible, and when he reaches a bounded facet it will be a feasible bounded facet (FBF) and this earns the player a point. However, any node might turn out to be an infeasibility trap in which case he has to restart from a low tree level. The ray matrix is a resource used by players to guide their next move to approach boundedness.

For a facet at a low tree level, there is generally a list of rays aligned to it, which proves that it is unbounded. To qualify as an FBF, a necessary (though not quite sufficient) condition is that all known rays are eliminated from the aligned rays list. So, it makes sense to choose each move such as to eliminate as many as possible rays from the aligned list.

A move from one node to the next is done by choosing a particular constraint hyperplane to add to the intersection list. Inspection of the corresponding column of the ray matrix reveals which rays will be eliminated by this move. The rays aligned with the current facet, are those which have zero elements in all columns belonging to its intersection list. Adding another column, will eliminate all rays with non-zero entries in the new column.

So, the count of non-zero entries in column j of the ray matrix can be taken as a **ray elimination score** for the corresponding constraint hyperplane. The score is a measure of the 'power' of column j to eliminate rays. This count runs over all rows q that belong to rays that are aligned with the current facet. Hence the table of scores needs to be updated after each move, as the elimination means that fewer rows are included in the aligned rays list when the facet level increases.

The scoring suggests a straightforward greedy algorithm for the search: at each node, select the branch with the highest score. This should not only lead to a bounded facet, but also do so in a low number of moves, as desired for finding a BFBF.

Unfortunately, there are several pitfalls in this simple strategy. One is that the highest score may lead to an infeasibility trap. So, it is necessary to test for feasibility before adding the highest scoring hyperplane, and if necessary, move down the scorecard until a feasible branch is found.

Another is that there is no guarantee that the greedy search will terminate at the lowest possible bounded level; it may well be that a different, less greedy, path will end up reaching a bounded level in fewer moves in total.

Further factors appear from considering the ray matrix. The ray matrix construction procedure described in Chapter 4, in particular Section 4.4, ensures that the first several rows of the ray matrix contain singleton rays. These by definition only has one non-zero entry in the slack columns of the matrix. If singleton ray q has its non-zero entry in column j, it can only be eliminated if column j is chosen in the branch decision. In other words, any bounded (i.e., ray free) facet must contain j in its intersection list. This makes j an *essential* hyperplane. Each singleton ray contributes a member to the set E of essential hyperplanes or **essentials**.

Since E will be a subset of the intersection list of any FBF (because otherwise the facet would have an aligned singleton ray) it means that the facet corresponding to E is itself an ancestor of all FBFs. It would therefore be a candidate to be designated as the progenitor facet. But it does not usually satisfy the progenitor requirement to be the *highest* level FBF ancestor facet.

An opposite case is presented by any ray free constraints as was used in Eq. (4.8). By definition, ray matrix columns that correspond to ray free constraints have consistently zero entries for all rays. So, they cannot eliminate any rays at all and are designated as **dispensables**.

All constraint hyperplanes that do not belong to either of the essentials or dispensable sets, are described as **possibles**. So, the intersection list of an FBF must contain all essentials, it cannot contain any dispensable, and consists of a subset of the possibles.

The simple greedy strategy is easily configured to exclude dispensables because they will have a consistent elimination score of zero. But essentials have a relatively low score value of one, so may fail to be included just based on their scores. To deal with this, before performing a greedy search the facet tree setup is tweaked in two respects:

- The root node is associated with the set E of essentials, instead of the empty list shown in Figure 5.2.
- The constraint number associated with each remaining node is picked from the list P of possibles only.

The reduced facet tree resulting from these modifications shrinks the search space and increases the chances for a successful greedy search considerably. For example, in the case of the Geobacter RSS with 138 constraints and 109 variables, there are 39 essentials and 12 dispensables. That only leaves 87 possibles out of 138 to constitute the search space, and the scoring narrows down the range of contributors even further.

6.3 A Randomised Greedy Search Heuristic

A randomised greedy search strategy that takes the various complicating factors into account combines the following steps:

(1) The ray matrix is used to partition the constraint hyperplanes into the non-intersecting subsets $\{E, P, D\}$, that is, essentials, possibles and dispensables.
(2) E is taken as the starting intersection list.
(3) Each node at the first level is allocated one member of P, and the branching nodes for its next level are only allocated subsequent members of P.
(4) Ray elimination scores are calculated for all possibles in P.
(5) At each node, the highest scoring branch is chosen as a candidate to be added to the intersection list accumulated so far.

(6) This prospective list is tested for feasibility, ancestry and boundedness.

  (a) If it is feasible, the candidate is accepted, else the next highest scoring branch is tried; exit the search if no branch remains.

  (b) If it is not a sub-facet of any known BFBF, the candidate is accepted, else the next highest scoring branch is tried; exit the search if no branch remains.

  (c) Else, ascend to the next level by accepting the candidate. If it is bounded, exit to step (7), else repeat from step (4).

(7) Backtrack the final FBF to its lowest bounded ancestor or BFBF.

(8) Add the current intersection list to the BFBF collection.

(9) Randomise the branching and repeat from step (2).

(10) Terminate the search when either a pre-determined number of BFBFs has been collected, or the number of consecutive failed searches exceeds an acceptable limit.

The minimal steps that are required for the search to deliver any FBF at all, for typical multidimensional polytopes, omit steps 6(b) and 7 to 10 from the stated strategy. The purpose of these additional steps is clarified by the following discussion.

The search can in principle fail if it terminates on an infeasible candidate. But perhaps surprisingly, that very rarely happens. In the Geobacter example, the search successfully finds an FBF at level 79, that is, starting from the 39-member list E it requires only 40 steps out of the possible 87 available in P to locate a solution. A purely random selection of 40 items from P, by contrast, can run over tens of thousands of trials without finding an FBF. This demonstrates the power of ray elimination scoring.

One drawback of the greedy search approach is that it traces a unique path through the facet tree and so only yields the same FBF if repeated. To address this, step 9 calls for introducing randomness.

In the Geobacter case, the original scoring was done using a ray matrix with 140 rays constructed as described in Chapter 4. During the search, four additional rays were found during boundedness testing and added to the ray matrix. Repeating the entire search with this slightly extended ray matrix, yields a new, different FBF also at level 79.

This suggests that the greedy search can be elaborated by repeating it with a randomly chosen subset of ray vectors in the ray matrix to get new solutions. That works — leaving out a small fraction (5%–10%) of randomly chosen rays gives scope for useful variation while ensuring that ray elimination scores remain realistic and yield a high search success rate. This is one way to perform the randomisation invoked in step 9.

In multiple randomised searches, an additional issue that arises is that the solutions found may be merely sub-facets of one another or even duplicates. Avoiding this is the purpose of step 6(b) in the procedure. It compares the prospective list with the intersection lists of all BFBFs collected so far. If it is a superset of any of those, it describes a sub-facet and the candidate is rejected similarly to the infeasible case. This does not protect against the opposite case that a previously found FBF may be a sub-facet of the current one. That possibility is avoided by the next elaboration, step 7.

As pointed out in the previous section, the simple greedy search has the shortcoming that it is quite likely to end up at an FBF that is not at base level. To remediate that, in step 7, each FBF is backtracked to its BFBF ancestor before being stored. The backtracking procedure is detailed in the next section. Finding BFBFs rather than just FBFs is the principal goal of the search, for the reasons given in Chapter 5.

A side effect of the backtrack step is that since a BFBF is by definition not descended from any bounded facet, it can never be a sub-facet of a newly found FBF. What is more, the sub-facet test ensures that BFBFs will never be duplicated. Whenever a search reaches a facet, even an unbounded one that is a sub-facet of any of the already collected BFBFs, this test terminates the search long before it even reaches another FBF that will backtrack to this known BFBF.

The price to pay for this guarantee is a higher search failure rate. As more and more BFBFs are collected, the probability that the tree traversal path fails because it will lead to a sub-facet of one of them, increases.

The result is that the randomised, backtracked greedy search is particularly effective in the beginning, but as the BFBF collection grows, the computational effort to add a new member to the collection steadily increases. On the upside, though, an increasing failure rate suggests that the majority of newly found FBFs are descended from known BFBFs, in other words, the set of BFBFs is approaching completeness.

That interpretation is only valid if the sampling can be assumed to cover the entire range of FBFs. But that seems doubtful when using ray subsets as described to introduce randomness. The slightly different scoring that results from a subset can be expected to give rise to a traversal path that still by and large covers the same region of the facet tree where the single maximally greedy solution was found.

To improve this and ensure wide coverage, while still exploiting the scoring to ensure a reasonable success rate, the randomised search sketched previously is replaced with a new strategy. This divides the search into two phases. In the first phase, relatively high scoring candidates are selected randomly from the possibles list P, and in the second phase, the remaining candidates are taken strictly according to score as before. The result is that departing from the root node, the main branches of the tree are selected rather randomly, but at higher levels the priority shifts to achieving boundedness.

To actually implement this strategy, the set P is first arranged in descending order of scores, each of which is an integer number by definition. Typically, only a few possibles have high scores but an increasing number of possibles share each lower score number. A fixed fraction ϕ of the range of scores is used to partition P. All members that belong to the upper fraction ϕ are shuffled randomly while the rest remain in descending order. The value chosen for the mixing fraction ϕ is a compromise between two considerations. Small values bring the search closer to the original version of maximum greed, and increase the success rate of an individual random search. In repeated searches, however, it gives a high probability of duplication so subsequent searches are prone to fail. A high ϕ value tends to cover a wider range of FBFs and reduces the probability of duplicate BFBFs on backtracking, at the expense of reducing the success rate of an individual search. Typical choices range from 70% to 90%. The midrange choice $\phi = 80\%$ is found to put around 40% of P in the group that is shuffled, and this gives a good combination of random coverage and moderate greediness in pursuing FBFs. Lower values collect FBFs more quickly in the beginning, whereas higher values are slower at the start but reach high FBF counts more quickly.

A further minor refinement of the algorithm is to notice that the inheritance of infeasibility can also be used for branching decisions. Suppose that at a node characterised by the intersection list f, it is found that adding candidate column j to that list will give an infeasible facet $\{f, j\}$. According to the 10-step procedure stipulated above it will be rejected and a different column, for example, g_1 selected to move to the next level. However, if at any subsequent higher level such as $\{f, g_1, g_2, ...\}$ there is again a choice to add column j, it will again give an infeasible facet because $\{f, g_1, g_2, ..., j\}$ is descended from $\{f, j\}$. So, during the traversal, a list of infeasible columns encountered on the path can be maintained and used to avoid infeasible branches without requiring an explicit feasibility test.

Finally, the dual criteria for stopping the search in step 10 are to be clarified. A random search can never be guaranteed to cover all facets. But, as shown below, for some purposes, it is sufficient to collect just a relatively small number of say 10 to 20 BFBFs as a basis for further calculations. In other cases, repeated failures to find new BFBFs has to be taken as an indicator that the search has essentially converged. A typical value for the acceptable failure limit is taken as 100 consecutive failures since the last success. This termination also takes care of the case that the chosen target number can never be reached, because the actual number of BFBFs for a particular polytope is smaller than the target number.

6.4 Backtracking and Traceback

The concepts behind backtracking and traceback were first introduced in Chapter 5 and backtracking is part of the random search as described in the last section. The traceback procedure has the goal of discovering the progenitor facet. The two procedures are broadly similar in that they aim to discover ancestors of known FBFs, but are quite different in the details of their execution. Of the two, backtracking is the simpler and is described first.

6.4.1 *Backtracking*

The aim of backtracking is to find the base level FBF, or BFBF, from which a given FBF is descended.

Traversing the facet tree to higher levels in a FBF search is accomplished by adding a constraint that belongs to subset P, to the intersection list of the current facet. Backtracking traverses in the opposite direction towards the root. Given the current facet specified by the intersection list $f = \{j_1, j_2, j_3, ...\}$, its parent is formed conversely by removing one of the j_i from the list.

In the backtrack procedure, the first step is to systematically test each j_i, except those that belong to the essential list E, for whether removing it gives a bounded parent facet. If none of them do, f is itself a BFBF.

Any constraint that does give a bounded parent, is referred to as a 'removable'. As each removable constraint is found, it is added to a subset r of f that collects removables. Now consider the ancestor of f formed by removing all members of r simultaneously:

$$a = Complement[f, r] \tag{6.1}$$

The facet a is the lowest level ancestor of f that can possibly be bounded.

To establish that, suppose for example that an earlier bounded ancestor b could be formed by removing in addition the constraint j_k from a. Then the facet c formed by removing just j_k from f would be a sub-facet of b. As the inheritance rules dictate that all sub-facets of a bounded facet are also bounded, this would mean that c is also bounded; but that contradicts the definition of r as containing all the removables. The same argument also rules out bounded ancestors at levels lower than b and formed by the removal of multiple constraints j_k.

If testing establishes that a is bounded, it is therefore the sought for BFBF. But usually, a will not be bounded. So, the second stage of the backtracking process is to start from the node corresponding to a, and then traverse the tree to higher levels, by using just the set r as the possible branching choices, until bounded facets are found. This traversal need go no higher than the level of the given facet f, as that is already known to be bounded. The limits on both the width and depth of the traversal means that it is quite practical to use the exhaustive tree traversal algorithm described in Section 6.6 .

Usually, this tree search will only yield a single bounded facet and that is the desired BFBF. Occasionally, the search might yield multiple BFBFs. In this case, they are all ancestors of f, albeit each on a different one of the multiple equivalent facet trees.

Computationally, backtracking is quite quick and can be used as a step that is applied to each FBF found by random search, even when the latter extends over thousands of repeated searches. As noted, the sub-facet check ensures that each backtrack that is executed delivers *new* BFBFs, and these are added to the list that keeps track of previous successes.

6.4.2 Traceback

To appreciate the benefits of traceback, it is informative to consider the results of randomised searches that are continued until a pre-determined number of FBFs have been sampled and backtracked. Figure 6.1 shows such trials for the *Geobacter* example.

In part (a) four trials that each sampled 1000 FBFs, are plotted together. In each trial, each randomly found FBF backtracks to one of four base levels as indicated by colour-coded traces. As more samples are collected, the number of distinct BFBFs at each level accumulates as shown in the plot. Because of the randomness of the sampling, the cumulative BFBF counts at each level are different in each trial at the same sample size, but maintain similar relative frequencies.

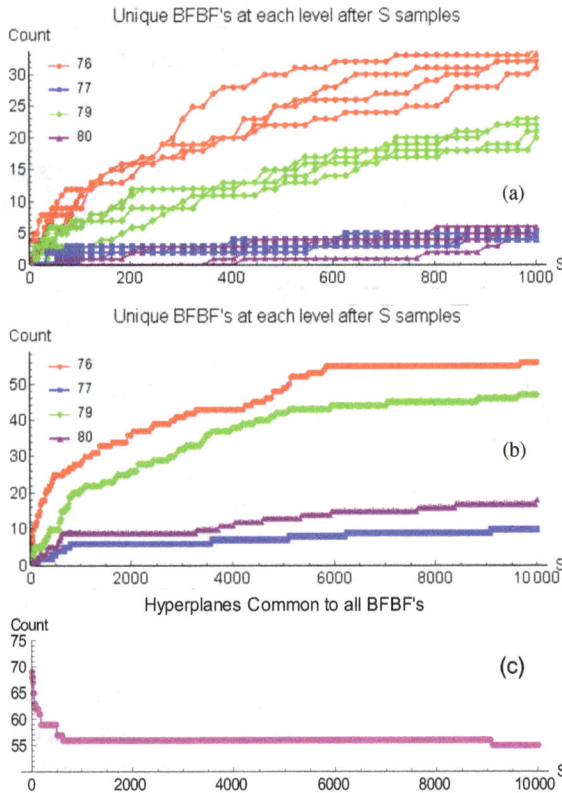

Figure 6.1   Accumulation of BFBFs with sample size, for each of the four colour-coded base levels. (a) Four trials of 1000 samples each, (b) One trial with 10,000 samples, (c) The number of hyperplanes that are common to all BFBFs at all levels, in the 10,000 sample trial.

The curves show a degree of saturation: new BFBFs are found more frequently in the beginning, but this tapers off as more FBFs are sampled. The result is that after sampling 1000 FBFs, only around 60 distinct BFBFs have been found in total.

Part (b) of the figure shows similar results for a single trial extending over 10,000 samples; a tenfold increase in the search effort only increases the total number of BFBFs found by approximately a factor of two. But only the trace for base level 76, which is found with highest frequency, appears to have converged to some extent. There is a law of diminishing returns that does not bode well for random sampling as an efficient way to find all BFBFs.

However, a striking observation from Figure 6.1 is that even though individual BFBFs accumulates slowly, from very early on the four distinct base levels at which they occur are unequivocally identified. As few as 20 to 30 samples are enough to establish these, and no further levels appear even after 10,000 samples.

A further significant observation is that there is a set of constraint hyperplanes that are common to all BFBFs found across all the distinct levels. This set of course includes the set E of 39 essentials that were shown in Section 6.2 to arise from singleton rays. But the set of **commons** is considerably larger than that, and as Figure 6.1 (c) shows has essentially converged to around 57 members after only about 500 samples were taken.

If there is a set C of commons shared by *all* BFBFs, not only the known ones, the tree node that corresponds to the set is by definition the progenitor node from which all BFBFs are descended. Sampling is quite successful in finding a first approximation to C. But the possibility always remains that some member of the approximate set will still be eliminated if sampling continues, and finds a BFBF that misses this member. In the example of Figure 6.1 (c), this still happened after about 9000 trials.

The idea behind the **traceback** procedure is to find the definitive set of commons in a more targeted way that eliminates or at least reduces any such uncertainty.

Traceback starts from the intersection list f that defines a BFBF node. Members of this list that do not belong to the essentials list E, are designated as non-essential and are candidates for removal in order to traverse to a lower tree level. The criterion for deciding whether a candidate j is in fact removable, is whether the parent reached by its removal has FBF descendants on a different branch from that of the known BFBF; in other words, a descendant that does not contain *j* in its intersection list. The traceback procedure cycles through all the candidates, testing each in turn for removability. If it succeeds, it proves that the starting BFBF is not an ancestor of all known FBFs since a counterexample was found. At the end of the cycle, deleting all removables from *f* results in a list *C* of hyperplanes that are common to all known FBFs, and hence defines their common ancestor node.

To guarantee that in fact all FBFs are descended from *C* as found from traceback, requires that the removability test is guaranteed to find any FBF descendants if they exist. That would be true, for example, of an exhaustive tree search. But for most cases, the test has to be based on the randomised greedy search which does not have this guarantee. In that case, *C* would only be an approximation to the true progenitor, and strategies to improve the approximation will be discussed in Section 6.5.

When the greedy search is used for removability testing, backtracking is excluded because it reduces the search success rate by avoiding FBFs descended from known BFBFs, whereas for the traceback search, all FBFs are relevant regardless of their mutual relationships. With this variation, the randomised greedy search has a high success rate and the traceback cycle is vastly more efficient to narrow down the set of commons than the simple repeated sampling strategy illustrated in Figure 6.1.

6.4.3 *Comparison Between Backtracking and Traceback*

Although both of the procedures function by testing a given facet intersection list for removable elements, they differ in several aspects:

- Backtrack starts from an FBF at any level, traceback from BFBFs.
- Backtracking terminates at a bounded facet, traceback at an unbounded facet.
- Backtracking is a strict procedure, not subject to randomisation, that is guaranteed to deliver the base level ancestor of the starting FBF.
- Traceback is more of a heuristic, that delivers an approximation to *C*, but depending on the quality of its removability test, it can miss a removable and terminate at a level higher than the true ancestor of all FBFs.

6.5 Traceback Determination of the Progenitor Hyperplane

For a reliable and efficient determination of the progenitor, simple traceback can be improved in two ways.

Firstly, the high initial success rate of the randomised, backtracked greedy search can be exploited to collect a small sample of BFBFs, rather than a single one, from which the traceback starting node is derived. As noted in discussing Figure 6.1, as few as 20 sample BFBFs are usually enough to collect members of the main families of FBFs as characterised by their different base levels. In setting up the commons list C, each sample BFBF is a counterexample for all hyperplanes not appearing on its intersection list, and that can hence be discarded. That leaves the hyperplanes shared by all sample BFBFs as a starting approximation for C. Avoiding the need to test discarded non-essential candidates during traceback may enable exhaustive testing of the members that remain in C. But even if not, it allows the computational effort of random testing to be concentrated on the cases where removability is harder to establish.

The second strategy is the repeated execution of the traceback cycle. Because of the random aspect of the greedy search, as well as the fact that elimination of other removables at the end of the previous cycle means that repeating the test for any given non-essential starts from a different node on the facet tree, it may well succeed to find an FBF on a subsequent pass.

The main problem to address here is how many traceback cycles are needed to gain confidence that all removables have been found and that the process has converged to the true progenitor G. Usually, the first few cycles each yield a few removables, but soon a stage is reached where no more removables are found over several consecutive cycles. Does this justify ending the search, or can a more quantitative stopping criterion be formulated? Figure 6.2 depicts the progression of the progenitor calculation for discussing this question:

Ray Mat		Possibles			Essential
BFBF's	Discard		Test 0		Essential
Cycle 1	Discard	Remove 1		Test 1	Essential
			...		
Cycle n	Discard		Remove n	Test n	Essential
			...		
Cycle ∞	Discard		Remove ∞		Progenitor

Figure 6.2 Progression of the progenitor calculation through multiple traceback cycles.

The ray matrix is used to distinguish between essential hyperplanes and possibles, of which a number D are discarded because they are not common to all of the BFBFs in the collected sample. At traceback cycle n, the number R_n of removables has been positively identified, leaving the remaining T_n possibles to be tested. Assuming that the removability test has a non-zero probability of success for every removable, after a sufficiently large number of cycles, all members of the removable subset $r = R_\infty$ will have been identified and only the hyperplanes that define the progenitor facet remain.

An exhaustive removability test will succeed with probability $p = 1$ for members of r and $p = 0$ for progenitor members. Sorting hyperplanes according to this probability and plotting, a step function is obtained.

For the randomised test, success in finding an FBF descendant depends on the relative number of FBF nodes among all nodes descendant from the starting node in the facet tree. The greedy search has a virtually perfect success rate for fairly abundant FBFs, but this decreases when they are relatively rare.

The result is that the success probability plot becomes a rounded step as illustrated in Figure 6.3. This behaviour is borne out in actual tests made, for example, for the Geobacter model, where 100 tests were performed for each of the possibles and success rates varied between 5% and 100% as a rounded step.

In statistical terminology, the randomised removability test has perfect *specificity* because its false positive rate $P(+ \mid !r)$ is zero. Here the positive outcome '+' means that an FBF descendant was found, while 'r' vs '!r' means that the hyperplane is removable vs not removable. On the other hand, the false negative rate $P (- \mid r)$ is not zero and its complement, the *power* of the test defined by $P (+ \mid r) = p$, is different for different hyperplanes. The plot in Figure 6.3 can be described as the power profile of the test.

In the plot, the horizontal axis shows sequence numbers in the hyperplane listing, ordered according to the power of the test. That means that the arrow lengths above the graph represents the numeric **count** of members belonging to each indicated subset of the possibles.

During the cycle repetition, removables connected with a large number of descendants are likely to be detected first, so the fraction of FBF descendants that remain unaccounted for decreases at a faster rate than the fraction of undetected removables. The following argument explores this idea in more quantitative terms.

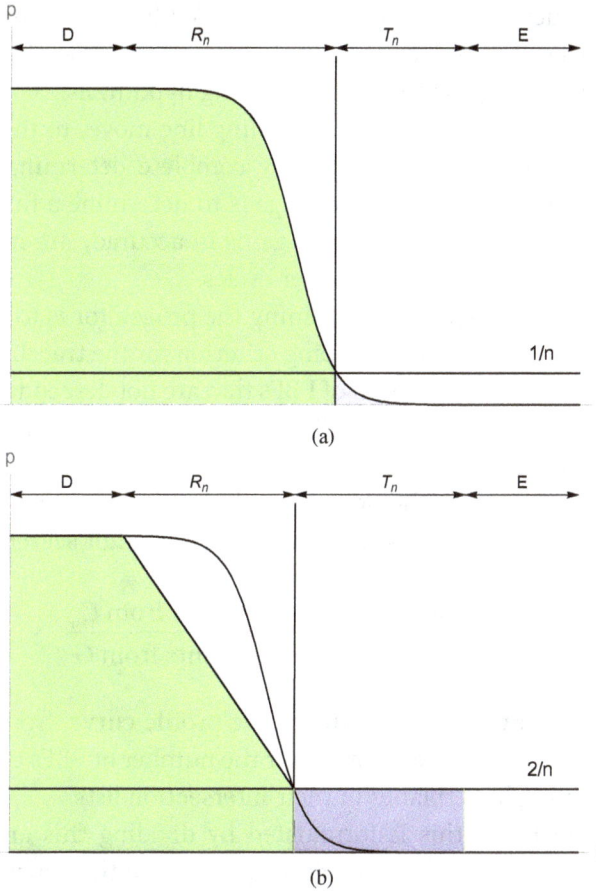

Figure 6.3 Schematic representation of the removability test profile: (a) Actual rounded step for power p versus sequence number i of hyperplane, (b) Simplified block version that conservatively estimates the missed fraction after n test cycles, based only on hyperplane counts D = discarded, R_n = removed, T_n = to be tested, E = essentials.

Testing a member of the removables set n times with a test of power p is expected to produce $n*p$ positive outcomes. As soon as a positive outcome is found the member is removed, so only those for which $p < 1/n$ are expected to survive after n traceback cycles. This is shown as a vertical line in Fig 6.3, separating proven removables from those still to be tested or known to be essential.

The random fluctuations around the statistical expectation that are inherent in the test, means that in practice, the division is better repre-sented as a diffuse vertical band rather than a line. The band is centred on the line shown. A removable element on this line is 50% likely to be identified as such in a given trial, and this likelihood increases towards the band edges. This band in effect defines a confidence interval, and setting its width at $2/n$ is a plausible working hypothesis.

As the value of n increases, the dividing line moves to the right, the band narrows and if taken to infinity complete determination of all removables would result. The challenge is to determine a finite n value that is sufficiently large that further gains in accuracy are not justified by the computational effort of further cycles.

Since the overall goal of determining the progenitor is to capture all FBF descendants, a plausible stopping criterion for the traceback cycle is to reduce the estimated fraction of FBFs that are not descended from the currently identified common set C, to an acceptably low value.

The connection between the power of the test and the number of descendants means that the power profile can be used to estimate the fraction F_n of missed FBFs, at cycle n of the progenitor search. Let:

$$F_n = 1 - \frac{\text{no. of FBF descendants from } C_n}{\text{no. of FBF descendants from } G} \tag{6.2}$$

It is proposed that the area under the profile curve, from zero to a chosen value i, is a reasonable proxy for the number of FBFs that contain all the remaining hyperplanes in their intersection lists.

The rationale for this is formulated by dividing this area up into vertical strips of unit width, each corresponding to the hyperplane with sequence number i. The removability tests similarly divide the FBFs into (possibly overlapping) subsets that omits one hyperplane at a time. The length of the strip is the success probability of the test, and is proportional to the number of qualifying FBFs, at least for cases where the length is smaller than one. Collecting all these subsets together, only the remaining hyperplanes are shared by all of them. Hence, the proposal as just stated.

Applying this to Eq. (6.2), at cycle n, the area under the curve up to $i = D + R_n$ can be taken as the descendant count for C_n and similarly the area under the entire curve as the count for G. Performing the subtraction, F_n reduces to the ratio of the blue area in Figure 6.3(a) to the sum of green and blue areas.

There are two factors neglected in this argument. First, in adding up the members of the omission subsets, there may be double counting. For example, an FBF that misses both i and j from its interaction list will be a member of the subsets connected to both i and j. This can be accounted for in the assumed proportionality constant between the count and the test power, and which anyway cancels out when taking the area ratio. The second factor is that strips of length 1, in general, underestimates the actual count as it only reflects the threshold count needed to guarantee success of the greedy search. This means that the area ratio is in fact a conservative estimate of the missing fraction, as its denominator is underestimated.

For practical application, there is also the difficulty that the detailed power profile curve is unknown, and as its shape depends on the way FBFs are distributed over the nodes of the facet tree, it can be expected to be different for each metabolic model.

But as the purpose here is to formulate a stopping criterion it is sufficient to establish just an upper limit to the fraction of missed FBFs. For this purpose Figure 6.3(b) shows a simplified block model, which only uses hyperplane counts to define its shape. By design, it overestimates the numerator of F_n (proportional to the blue area) and underestimates the denominator (proportional to the sum of green and blue areas) so as to give an upper limit rather than an actual estimate of the fraction.

The over/underestimates result from the following assumptions that underlie the approximation:

- The dividing line between confirmed removables, R_n, and hyperplanes to be tested further, T_n, is taken at a power value $2/n$ rather than the expectation value $1/n$. This takes into account the probabilistically broadened band, and in effect moves the dividing line to the left so increasing the blue area without affecting the combined area.
- The linear power decrease underestimates the green surface area.
- The rectangular blue block implies that a fixed power $2/n$ is assumed for members of T_n. This overestimates the blue area, and does so to a larger extent than the small increase it that it also produces for the combined area.
- Assuming the power value to represent the number of FBF descendants, neglects the fact that the power saturates at a value 1 for FBF ratios higher than a threshold value. So, the green area is itself an underestimate of the FBF counts from the region at the top of the step.
- The perfect specificity of the test introduces an asymmetry: all hyperplanes identified by the green area are certain to be removable,

whereas only a fraction of those associated with the blue area are removable.

This fraction is again conservatively assumed to be $2/n$.

Using the block approximation, the upper limit for the estimated missed fraction at traceback cycle n is as follows:

$$F_n \leq \frac{\frac{4}{n^2}T_n}{\frac{2}{n}(D+R_n)+\frac{1}{2}(D+R_n+D)\left(1-\frac{2}{n}\right)+\frac{4}{n^2}T_n}$$

$$= \frac{8T_n}{n^2(2D+R_n)+2nR_n+8T_n} \tag{6.3}$$

In practical application, the traceback cycles are terminated when the right-hand side formula of Eq. (6.3) drops below a fixed value of 0.005. This typically requires $n = 10$ to 20 cycles, so determining a progenitor that is the common ancestor of more than 99.5% of all FBFs.

In the case of the Geobacter example, 14 cycles were needed to reach this level and results in a progenitor at level 53. The 10,000 random searches shown in Figure 6.1 required more than 10 times the computational effort, and even then stopped at level 54. This demonstrates that the traceback approach is more efficient by an order of magnitude or more, in addition to giving a quantitative measure of its accuracy. In fact, for the Geobacter case, numerous repeats of the progenitor calculation always produced the same 53-member facet, suggesting that it actually captured 100% of all FBFs, even though this is impossible to prove for any random search.

As outlined in Chapter 5, Section 5, determination of the progenitor is usually followed by projecting the RSS to the progenitor hyperplane and then repeating the FBF search in this lower dimensional subspace.

6.6　Setting up the Facet Tree for Systematic Traversal

As an alternative to the elaborated greedy search, it is in some cases possible to perform a BFBF search by systematic traversal of the facet tree. This will usually only be practical in relatively low, double digit dimension counts, and even then, require every possible simplification strategy. Some of these were already indicated, such as the use of inheritance rules, but this section aims to set out the strategy in more detail.

A key part of this is similar to the well-known 'branch-and-bound' algorithm used, for example, for integer domain LP calculations, to optimise an objective by dynamically pruning branches that will not yield solutions outside of known bounds. Here, the pruning is done where further traversal will not deliver a new BFBF. This is decided on the grounds of various stopping conditions listed below.

6.6.1 *Structuring the Tree and the Traversal*

But first, it is important to optimally structure the facet tree itself. Because of the tree asymmetry, it is advantageous at each tree level to traverse the branches from right to left on the tree layout in Figure 5.2. This ensures that BFBFs will be encountered first, before their descendants, as those are located below and to the left. Also, it exploits the fact that the tree depth is lowest on the right, allowing more main branches to be traversed for a given computational effort.

A second structural aspect is the choice of the root node. If a tree search is done for progenitor determination, the node associated with the essentials list E is chosen. As explained previously for the case of the greedy search heuristic, all FBFs are descended from this node and the discussion in Chapter 5, Section 2, shows that this choice merely implies selecting a particular subset out of the diversity of equivalent facet trees. However, if the progenitor has already been determined using the greedy search with traceback, this would normally be followed by projection to the progenitor hyperplane in which case all its members G, which includes E, have already been eliminated and the root node has an empty intersection list. In some cases where no ray could be found that is orthogonal to the progenitor, projection is impossible and then the root node intersection list is taken as G.

Further choice among the diverse facet trees is possible by fixing the sequence in which possibles are taken to define tree branches. As described before, the asymmetric facet tree conserves the chosen ordering. The optimal choice would put all BFBFs in the upper right of the tree diagram, so that they can be reached by traversal of the shallowest main branches. A good ordering of possibles can be crucial for a large-dimensional polytope, as the exponential increase of nodes towards the left can rule out traversal of all main branches and fail to find any FBFs within a practical computing time.

An obvious way to arrange the set P of possibles is in ascending order of their ray elimination scores that were used for greedy searches. In this way, the most likely candidates are picked up first during the right-to-left traversal. This works quite well for smaller facet trees, but not perfectly, because as pointed out earlier, some hyperplanes with high scores remain absent from BFBFs. For larger trees, better results are obtained by first collecting a BFBF sample by greedy search, and then using the frequency count of each hyperplane over this collection as its score. Further light will be thrown on how these methods of ordering compare, in the example discussed in Section 6.8.

A final consideration is that most tree searches may be restricted to cover only a limited range of tree levels.

This is in particular true of the backtrack search, where the tree search starts at the level of the node obtained by omitting all removables, and ends at the level of the known FBF.

During traceback, the search by design only traverses branches outside the range of known BFBFs, so there is no hard limit to how high it should go. But as demonstrated in Figure 5.1, FBFs generally occur over a limited range of levels, and BFBFs mostly in the lower part of this range. This is also borne out in the example shown in Figure 6.1 where all BFBFs are confined to a narrow range of levels. So, it makes sense to limit a traceback tree search to only a few levels below the known BFBF levels from which the traceback was started.

Finally, a similar argument applies even to the general BFBF search; whereas neither a lower nor an upper level can be determined absolutely, plausible limits can be put for levels where BFBFs are expected, based on the preliminary BFBF sample, as further explored in Section 6.8.

In all three cases, BFBF tree searches can be limited to those that terminate between minimal and maximal levels, and the limiting levels can be set a priori.

6.6.2 *Subtree Viability*

The ray matrix also provides one of the main tools for dynamic pruning during traversal. It allows a decision of whether a tree branch is **non-viable**, in the sense that there are no bounded facets descended from it.

This is explained by noting that since traversal maintains the ordering of possibles, only those with numbers that appear *after* that of a starting node in the fixed sequence of possibles, can be picked up in a traversal of its descendant branches. If the sum of ray elimination scores of these subsequent possibles is less than the number of known rays in the facet, they cannot all be eliminated so the starting node is non-viable.

In fact, the viability criterion can be made sharper by checking if the specific subset of rays aligned with a particular facet, can be eliminated by the subset of possibles that are available to be collected in traversing its subtree.

Recall that the intersection list f associated with a given node V in the facet tree, defines a subset of columns in the slack or overlap section of the ray matrix. The known rays that are aligned with the facet corresponding to V, are represented by all rows of the ray matrix that have uniformly zero entries in the columns defined by f. Encapsulate this information by constructing a binary vector *InFacet* that has a unit entry for each ray matrix row that contains an aligned vector, and zero otherwise.

For every node V at level k of the facet tree, the last entry in the facet intersection list, namely m_k, identifies its constraint label in the tree representation. Given the ordered sequence $P = \{p_i\}$ of possibles that apply to the root node of the tree, m_k is by construction a member of this set, say $m_k = p_j$. Then the set that are available at node V to construct a subtree, is the subset $P_V = \{p_i : i > j\}$. Again, P_V defines a subset of ray matrix columns, and which does not overlap with f. As before, introduce a binary vector *NonElim* that has a unit entry for each ray matrix row that has uniformly zero entries in the columns that belong to P_V, and zero otherwise. This vector indicates all rays that cannot be eliminated by traversing the subtree emanating from V.

The viability of node V can now be formally calculated as a binary value:

$$viable = 1 - Sign[InFacet \cdot NonElim] \qquad (6.4)$$

It is safe to assume that the starting node of a tree search is always viable (if not, no FBF exists and the search is abandoned). Notice that if node V is not viable, all nodes V' at the same level and to its right on the tree diagram are also non-viable. This is because the set P_V' that belongs to each of these is a subset of those belonging to V, and so even more restricted in their ability to remove rays.

Because of this feature, it is more efficient to apply the viability test in a forward-looking fashion. Having calculated the *InFacet* vector for vertex V, the vector *NonElim* is calculated for each of its descendant vertices in ascending order of their index value, that is, from left to right in terms of the facet tree (i.e., opposite to the traversal direction), and this stops at the first descendant that is non-viable. This and all further branches can then be pruned away as non-viable.

Note that the viability test relies on the set of known rays in the ray matrix, which is generally not complete. If the test fails, it means that there are one or more known rays with zero entries for all columns that belong to either f or P_V. That fact is not changed by whatever further unknown rays exist. So, if a node is found non-viable, that result is conclusive despite the ray matrix being incomplete.

On the other hand, if a node is found viable according to Eq. (6.4), the result could be overturned by finding another ray that does have zero entries in all f and P_V columns.

A relevant example of this arises if in calculating viability, it is found that *InFacet* is the zero vector. This means that V defines a ray free facet and has to be LP tested for boundedness. If boundedness is confirmed, V is an FBF and traversal terminates. Boundedness can only fail by detecting a new ray with zero entries in the f columns, and if so, the new row is added to the ray matrix.

In this case, viability testing for V should to be repeated with this extended ray matrix, and in principle, nodes that were previously found viable may need retesting too. But as nonviability is clearly an inherited attribute, any missed nonviability will be detected at the next tree level anyway so there is no harm done in skipping the retesting.

The conclusion is that an incomplete ray matrix may lead to occasional missed pruning opportunities, but this is mostly remedied by dynamically updating the ray matrix during traversal. Most importantly, viability-based subtree pruning is safe as it will not incorrectly prune away viable subtrees.

6.7 A Tree Traversal Heuristic

As for the greedy search, the starting point is the ray matrix that is used to partition hyperplanes into sets E and P, allocate ray elimination scores to the members of P, and sort them accordingly.

The search starts at the root node, which is either E, or in case the progenitor has already been calculated, it may be G or the empty set. For tree searches done for backtracking or traceback, the starting node is fixed by the procedure.

At each node V, there is a subset of P that are available to be added to its facet list in constructing its subtree. This candidate subset starts as the set P_V introduced in the previous section, but is further reduced by pruning away non-viable branches, depth pruning or traversed branches.

In essence, tree traversal is a recursive process that starts at the chosen node, and either stops or proceeds to the rightmost unexplored node in the next higher tree level as determined by the following traversal heuristic. If stopped, recursion reverts the search to the last lower-level node while removing the stopped branch from its set of candidate possibles.

1. Test the facet associated with the starting node for *feasibility* and *boundedness*.
    1.1. If feasible and bounded, it is the only BFBF as all remaining FBFs descend from it. Append it to the running BFBF list and end the search.
    1.2. If bounded but infeasible, end the search as all its sub-facets are infeasible.
    1.3. If feasible but unbounded, proceed.
2. Test all descendant nodes, in ascending sequence, for *viability*.
    Stop at the first one found non-viable and remove it and all subsequent elements from the candidate possibles list available to V.
3. Check the *depth* of all the remaining descendant branches, and also remove any that do not penetrate to the minimal preset tree level.
4. If the remaining candidate list is empty, revert to previous level and repeat from step 4.
    If reversion is impossible because the current node is the starting node, exit the search.
    Traverse to the next level by adding the last entry on the candidates list to the current facet list.
5. Test if the node is a *sub-facet* of any facet on either:
    5.1. The currently known list of infeasibles
    5.2. The currently known list of BFBFs

If so, revert to the previous level and repeat from step 4.
6.   Test the facet for *feasibility* and *boundedness*.
   6.1. If feasible and bounded, append it to the running BFBF list, revert and repeat from step 4.
   6.2. If infeasible, append it to the running infeasibles list, revert and repeat from step 4.
   6.3. If feasible but unbounded, proceed.
7.   Check if the preset maximal level has been reached, and if so revert and repeat from step 4.
8.   Check if the preset maximal number of BFBFs or the maximal number of tree nodes to be visited has been reached; if so, end the search.
9.   Repeat from step 4 without reversion.

The traversal heuristic has been designed to execute quicker tests first before the more intensive combined feasibility and boundedness test.

In principle, this procedure will systematically traverse the entire facet tree, except for the subtrees that were pruned out, and find all BFBFs that exist. In practice, that is often not tractable because of the exponential increase in node count for each subsequent main tree branch. Setting a maximal tree level helps to ameliorate this and as an additional safeguard, the heuristic provides for setting predetermined maxima for both the total number of nodes to be visited, and the maximal number of BFBFs to be collected.

For perspective, in the case of the Geobacter example, it is found that roughly 100 nodes can be traversed per second and taking the maximal count as 500,000, constrains tree searches to a time scale of hours. This limits the number of main branches of the facet tree that are traversed to a range of 10 to 15 for models with a complexity similar to the Geobacter model. That is only a fraction of the 45 branches that remained in that model after pruning.

The limitation to around 10 main branches may not be a problem if the reordering of possibles succeeded in placing all BFBFs far enough to the right in the facet tree. The case discussed in the next section shows that unfortunately this cannot be relied upon to happen.

This means that only for modest facet trees can a traversal search really be expected to be comprehensive. In larger cases, similar to the random search, it simply has to be abandoned because of a law of diminishing returns on computational effort.

Recognising this fact, the traversal heuristic allows for a preset maximum number of BFBFs to be found. This maximum target is typically set at 500 or 1000 for both the tree and greedy searches.

Further justification for setting such a maximum is the anecdotal observation that the tangential capping radii determined by the set of BFBFs are quite insensitive to the choice of this maximum. After all, the main purpose of finding BFBFs is to calculate capping constraints. In trials on the Geobacter model, virtually the same capping radii were found for as few as 100 BFBFs as for a trial that persevered until 2000 BFBFs were found. It would seem that obtaining just a representative sample is more important than determining the comprehensive set of BFBFs.

6.8 Choosing Between Greedy Search and Tree Traversal

For a limited search such as for backtracking or for a small facet tree, the tree search is clearly preferred because it is comprehensive. However, for large searches, the law of diminishing returns affects the tree search even more severely than the greedy random search.

That is illustrated in a case study for a model consisting of 64 constraints in 44 dimensions, which is fairly representative of medium-size models after projection to the progenitor facet. Here, a tree search found 959 BFBFs in the first 12 hours, and when terminated after a total of 23 hours having covered about 7 million facets, only increased the FBF count to 1004. By contrast, one random greedy search found 906 BFBFs in 15 minutes, whereas another that was given a target count of 2000 BFBFs achieved this goal in 3.2 hours after visiting 2.1 million facets.

A more detailed comparison between the two approaches in that case study is quite instructive. Various random searches established that out of the 64 constraint hyperplanes, only 45 participated in any of the BFBFs. For each of these, the frequency of its occurrence over a typical set of 906 BFBFs from one such search, was calculated. Then, a tree search was done in which the set of possibles was restricted to the 45 participants, arranged in ascending order of their frequencies. This

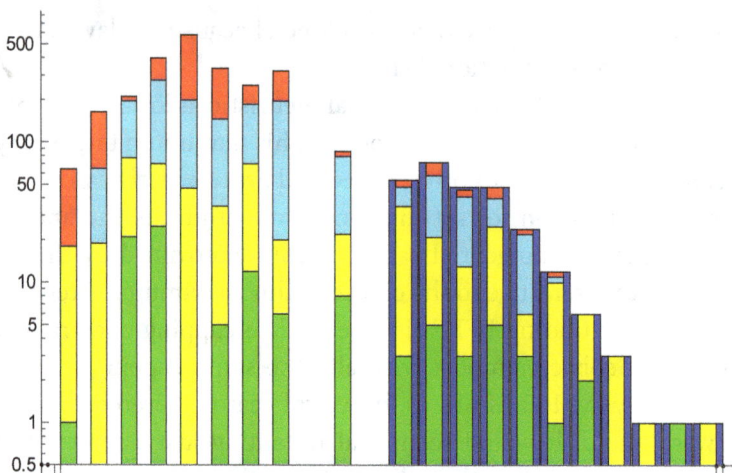

Figure 6.4   Allocation of BFBFs found in various searches, to the 22 viable tree branches in an example. Blue bars represent the 11 branches that were traversed (right to left) in a tree search. Superimposed on these are coloured bars for BFBF sample sets collected in a series of random searches with increasing sample sizes: Green – 100; Yellow – 500; cyan – 1593 and Red – 2685 samples.

allowed the tree search to prune away the first 23 branches as non-viable and traverse the next 11 branches comprehensively.

Next, a series of random searches were conducted that each terminated at an increasing total BFBF count. Each member of the BFBF set produced in each random search, was allocated to its appropriate branch of the tree that was partially traversed in the tree search.

The results are presented graphically in Figure 6.4. The blue bars represent the total number of BFBFs found by the tree search, for each of the first 11 branches (out of a total of 22 viable branches) that it traversed.

It is striking that the majority of BFBFs found in each random search, are in fact located on subsequent main branches of the tree (note the logarithmic scaling in the graph). The tree search only delivered a total of 270 BFBFs, while the longest random search found 2685. The tree search visited almost a million tree nodes and needed about 1.4 hours to find the 270 BFBFs, while random searches routinely find more than 300 in the first 5 minutes. On the other hand, the tree search did cover its branches comprehensively; for those branches, the series of random searches yield increasing numbers of BFBFs approaching, but not exceeding, the tree search counts. In total, the 2685 randomly found

BFBFs include 268 out of the 270 (or 99.3%) on the branches covered by the tree search. This suggests that the vast majority of all BFBFs have been found in the largest random search, albeit at the cost of visiting 78.5 million nodes and 160 hours of computing.

The fact that the tree search could only access about 10% of all BFBFs in the branches that it could realistically cover, suggests that the branch ordering used for the search did not succeed in shifting all BFBFs to the rightmost part of the tree as it was intended to do. Even so, the importance of ordering is demonstrated by noting that a randomly chosen order typically results in all BFBFs being located on just the leftmost two branches. So, the shift was at least partly successful. In fact, arranging the branches according to their ray elimination scores only succeeds in spreading BFBFs over a total of 11 viable branches, instead of the 22 viable branches of the frequency ordering. With these 11 branches, even the first viable branch is already so deep that its comprehensive traversal fails to find a single BFBF in a reasonable timespan.

It is noted from Figure 6.4 that even the relatively modest random search for 500 (or about 20%) of all BFBFs, successfully identifies all BFBF levels and participating tree branches, and hence provides an estimate of relative frequencies that could be used to establish the branch ordering.

In fact, even the very small samples (up to 20 BFBFs) that are collected to establish the progenitor hyperplane, are found to yield adequate frequencies to establish a node ordering for a viable tree search. Even for searches of trees that are small enough to be traversed completely, so that the branch ordering does not matter in principle, it is still vastly more efficient to use the ordering based on frequencies from the progenitor sampling rather than elimination score ordering. This is because frequency scoring allows the first BFBFs to be encountered sooner, and hence used to prune away large parts of the facet tree.

Considering the comparison, both the randomised greedy and comprehensive searches need to contend with the fact that only a vanishingly small fraction of facets are BFBFs for a polytope in 44 dimensions. The tree search deals with this by using every available trick to avoid searching non-productive branches and to stop at the base level FBF, that is, the BFBF. In the random search, the greedy aspect steers it towards FBFs, but it then adopts the apparently wasteful strategy of tracing each FBF it finds, back to its BFBF ancestor, and then rejecting this if it duplicates one already found. By the end of the greedy search, more than 99% of all

BFBFs found are rejected. Nevertheless, it turns out that this is still faster than the comprehensive search by an order of magnitude.

The main lesson learned from this example, is that the randomised greedy search is so much more efficient than a tree search, that the latter should only be used when full traversal of the relevant part of the facet tree is realistically possible.

That raises the issue of how to predict the scope of the required traversal to make this decision. An obvious criterion is the total tree size, which is easily calculated using Eq. (5.1) as containing 2^p nodes, where p = total number of possibles. Taking the number of nodes that can realistically be traversed as a number in the range 10^5 to 10^6, this would limit tree searches to cases with $p \leq 20$. That is unnecessarily pessimistic, as it ignores the large number of nodes avoided by all the various pruning strategies. More realistically, p can be reduced by only including the subset of possibles that actually participate in BFBFs and restricting searches to the band of tree levels where BFBFs are actually found. These two factors are straightforward to estimate, while the further reductions because of inheritance rules are more difficult.

One approach to estimate relevant tree levels is to find the minimum and maximum levels at which all rays could be eliminated, based on the ray matrix elimination scores. However, these limits are typically too wide to be useful.

The decision about which BFBF search method to use, in fact needs to be made twice: first, when determining the progenitor hyperplane, and again later for the full search. Progenitor determination only needs a representative sample, so the random sampling method is appropriate unless the node count is so small that a comprehensive search can be completed in negligible time. A pragmatic choice is to take the limit at $p = 14$, corresponding to a node count of $2^{14} = 16,384$ for which the optimised tree search should complete in a matter of seconds.

At the stage where a full search is attempted, the availability of the progenitor sample can be used to make a better estimate. A typical sample size is in the range of 10 to 20 BFBFs. These generally only represent the most common BFBF levels and their hyperplane participants. Anecdotally, it is found that even a single BFBF already contains about half of all participants. So, to estimate the participant count, a multiplicative correction factor f is applied to the number observed in the sample, and is obtained from the following formula that smoothly interpolates between a value around 2 for a single BFBF, and 1 at infinite sample size:

$$\tilde{p} = f \cdot p \tag{6.5}$$

$$f = 1 \Big/ \Big(1 - 0.4 e^{-0.055 \cdot sample\ size}\Big) \tag{6.6}$$

The numerical parameters were chosen to fit trial values with varying sample sizes and a participant number well established by large random BFBF searches. The empirical formula is not claimed to be particularly accurate, nor does it have to be as it is merely used in the decision about which type of search is done, and has no effect on the actual results of either search.

For the reasons explained in Chapter 5, FBFs fall in quite a narrow range of levels, and BFBFs even more so. This was also illustrated in the example of Figure 6.1. If the number of nodes in the most populous of these levels can feasibly be visited in a tree search, inclusion of the few remaining levels does not make a material difference.

So, the decision criterion is taken as the node count in the most populous level, for a facet tree with $\tilde{p}$ participating hyperplanes (i.e., main branches). The number of nodes at level k was shown in Chapter 5 to be the binomial coefficient $^{p}C_{k}$, which reaches its maximum at $k = \tilde{p}/2$. Hence, the most populous BFBF level is taken as the level found in the sample that falls closest to the value $\tilde{p}/2$.

If the corresponding binomial coefficient exceeds the chosen maximal number of nodes that can be processed, the decision is taken to use a randomised greedy search instead of a tree search for BFBFs.

Applying these considerations to the 64 × 44 example discussed previously, a run with sample size = 15 produced a total of $p = 38$ participants. The correction calculated by Eq. (6.6) extrapolates that to the estimated $\tilde{p} = 46$. Recall that the actual number of participants from large samples was 45. The sample contains BFBFs at levels 21, 24 and 25, of which level 24 is closest to $\tilde{p}/2 = 23$. The node count estimate is $^{46}C_{24} = 8 \times 10^{12}$, which rules out a tree search in the case of a search assumed to be limited to 10^{6} nodes.

This is expected to be an overestimate because of neglect of further pruning, somewhat offset by not accounting for the nodes at lower tree levels that are visited on the way towards the terminal nodes at each BFBF level.

As an example, the tree search illustrated in Figure 6.4 stopped at main branch 34 (23 non-viable and 11 branches searched comprehensively). Taking as above the representative BFBF level as 24, on this truncated tree, there would be an estimated $^{34}C_{24} = 10^8$ nodes on this level, but the comprehensive search in fact only needed to visit around 10^6 nodes in total.

It follows that the proposed criterion tends to err on the side of choosing a random search; if it chooses a tree search, this should complete safely within the acceptable time limit.

Note that a simple tree size criterion would have been even more pessimistic, in the case of this example, by a factor of well over 10^5.

A final use of the BFBF sample is to set minimal and maximal levels for the tree search. Anecdotally, no more than six BFBF levels have been found in any model investigated. A conservative estimate is to search all levels within two levels below or above the minimal and maximal levels that occur in the sample.

Chapter *7*

Constructing and Characterising the Solution Space Kernel

7.1 Tangent Capping of the Solution Space

The stages set out in the previous chapters used coincidence capping and the consequent projective transformations to remove the majority of rays and dimensions from the specification of the solution space of a Flux Balance Analysis (FBA) model. For most models some rays still remain, even after projection to the progenitor space. This means that the SS remains a partially open polytope. To construct the desired compact kernel, tangent capping still needs to be executed. The determination of the set of base level feasible bounded facets (BFBFs), covered in the previous chapter, is the major step required to make this possible.

To construct a capping hyperplane, its orientation (i.e., a unit vector ĉ along the direction of the capping ray) is chosen first as further described below. Then, the minimum capping radius R needs to be determined, such that the capping hyperplane only intersects any of the BFBF's tangentially, for example, at a single vertex point or at an edge.

Note that the capped solution space, the Solution Space Kernel (SSK), is not uniquely defined. Various choices can be made for the capping rays giving rise to distinct sets of capping hyperplanes. Capping is, after all, merely a partitioning of the SS, and different ways of partitioning it may all satisfy the requirements spelled out in the discussion of Eq. (1.12). One possible aim to pursue in choosing capping rays is to minimise the number of additional constraints needed to fully close the SSK. But, as the successful dimension reduction already leads to quite a small set of constraints that specify the open Solution Space (SS), this seems less important than achieving a compact shape that reflects the shape dictated by the bounded facets of the SS. Recalling that the SSK contains all the vertices of the open SS polytope, the term 'compact' can be taken in this context to mean that any additional vertices introduced by capping hyperplane intersections should remain as close to the SS vertices as is possible for a convex polytope. The need for this compactness goal and a strategy to achieve it is further elucidated below.

To formulate the calculation of a capping radius as a Linear Programming (LP) problem, consider a point X that belongs to a particular bounded facet F. The intersection list f of this facet identifies the subset of polytope constraints that are satisfied as equalities for any point located on the hyperplane on which F lies. The boundaries of F are formed by the intersects of the remaining constraint hyperplanes that define the polytope, so X satisfies all such constraints as well. In order for the capping hyperplane not to bisect F, it has to be placed at a radius R_F that is large enough that all points X are on the same side of the capping plane as the coordinate origin, so $\hat{c}.X \le R_F$ for all points X. This is achieved by maximising $\hat{c}.X$ and setting R_F to this maximum value. Repeating this LP calculation for each F that belongs the BFBF set, the final capping radius R is taken as the largest R_F value.

With this strategy, the capping hyperplane will usually only intersect a single facet, the one that yielded the largest R_F, and may intersect it only at one vertex point. In principle it should be possible to improve on this. In three dimensions (3D), for example, one can imagine fixing the capping plane to this vertex, and then rotate it about this point to intersect a second facet at one of its vertices, and finally rotate the capping plane about the axis formed by the first two until it intersects another facet at a third vertex. Generally, in N dimensions, there are N degrees of freedom and so it should be possible to find capping hyperplanes that tangentially intersect up to N facets simultaneously. Numerical procedures to implement this were tried but found to suffer from convergence problems and were not further pursued.

One method for choosing $\hat{c}$ was discussed in Chapter 4, Section 6, as an 'interior' ray direction that is intermediate between peripheral ray directions. As mentioned there, such a direction may cap several ray directions simultaneously, but not necessarily all and it may be necessary to repeat the process and find several capping ray directions before the solution space is fully closed.

In practice, this method is quite successful in finding a small capping ray set, but the resulting kernel polytope sometimes still has a large extension in some directions. The reason for this can be understood by considering the example shown in Figure 4.4. Suppose that either of the blue arrows in the figure was chosen as a capping ray rather than the red arrow. The resulting polytope would still be closed, but the capping plane intersects one of the semi-infinite sides of the open polytope quite far away from the origin and creates a spurious 'spike'. This problem arises whenever there is a large obtuse angle between the capping ray and the constraint vector of an unbounded polytope facet, giving two nearly parallel constraint lines.

In the example, choosing the red arrow avoids this and gives a more compact kernel; so does using both blue arrows to give a kernel bounded by two intersecting capping lines, and the most compact kernel is obtained by combining all three ray vectors. This generally increases the total number of constraints that define the kernel, but the example shows that in fact some of the constraints at the vertex where the capping hyperplane intersects the SS may become redundant and can be eliminated, depending on the exact orientation of the bounded facets.

The principles illustrated here carry over to higher dimensions. Each peripheral ray is by construction orthogonal to the constraint vector of at least one unbounded facet. Including them all in the capping ray set ensures that no unbounded facet will be capped by a nearly parallel constraint hyperplane, and so avoids spurious spikes. So, the general strategy chosen is to combine the full sets of interior and peripheral rays as candidate capping rays and find a capping radius for each.

Based on this discussion, the tangent capping procedure consists of the following steps:

(1) Assume that the partially unbounded SS polytope has the half-space specification

$$C \cdot x \leq V \tag{7.1}$$

and that its bounded facets are known as a set B of intersection lists of constraint hyperplanes identified by their row index m in C:

$$B = \{f_i\}; \qquad f_i = \{m_1, m_2, \dots m_k\} \tag{7.2}$$

(2) Calculate the set of interior rays that guarantees polytope closure, and combine this with the set of peripheral rays embodied in the ray matrix to give a candidate set of capping ray directions $\{\hat{c}_n\}$.

(3) If the SS is a simple cone, B will be an empty set. In this case, allocate a pre-assigned default capping radius to all capping ray directions and exit the procedure.

Else, the SS is a facetted cone and B has one or more members.

(4)   Select ray n, starting with the first ray, $\hat{c}_1$.

(5)   For facet i in B, set up the modified constraint set $\mathbb{C}$ by changing rows m_i, $i = 1, \ldots k$ into equalities. Then, use LP with the objective vector $\hat{c}_n$ to maximise $\hat{c}_n.x$ subject to $\mathbb{C} \cdot f \leq V$ and set R_{ni} to the maximum obtained.

(6)   Repeat step 5 for all members of B, then set $R_n = \text{Max}[R_{ni}]$.

(7)   Repeat steps 4–6 for all rays to collect a full set of capping radii.

(8)   Set up an extended constraint matrix C_k by inserting all capping ray vectors as additional rows into C, and an extended values vector V_k by inserting the capping radii in corresponding order into V. The compact kernel of the SS is the bounded polytope defined by the extended constraint set

$$C_k \cdot x \leq V_k \tag{7.3}$$

(9)   Check V_k for any negative entries indicating that the coordinate origin has been excluded by capping. If so, restore the origin to an interior point by using the constrained linear ordered differences (CLOD) recentering as described in Chapter 3, Section 2.2, and adjust the capping radii to the new centre.

(10) Check that the maximum capping radius significantly exceeds the fixed value tolerance. If not, the kernel is essentially a point and treat the SSK as a simple cone by performing step 3.

(11) Eliminate any redundant capping constraints as described in Chapter 3, Section 3.

(12) Find the maximal inscribed hypersphere using the method described in Chapter 3, Section 2.3, and move the origin to its centre.

(13) Check if the maximal inscribed radius is negligible. If so, improve centering by moving consecutively along each capping ray direction to its midpoint.

There are several details of the procedure that deserve further clarification.

The first is the distinction between simple and facetted cones. A *simple cone* is formed when all constraint hyperplanes intersect in a single point, namely the cone apex. An example of this is the stoichiometry constraints in Eq. (1.1), that is, $S \cdot x = 0$, which underlies all FBA models. This is solved by the flux vector $x = 0$ that defines the coordinate origin.

So, this single point lies at the intersection of all constraint planes since it satisfies all the constraints simultaneously, defining the origin as the apex.

This trivial zero flux solution is ruled out in most models by the further constraint of a finite objective value. That introduces bounded facets and hence gives rise to a *facetted cone*. However, it can happen that after the various dimension-reducing projections, the constraints that remain in the progenitor hyperplane once more describe a simple cone, although with an apex representing non-zero flux.

It is quite straightforward to establish whether this is the case without embarking on a feasible bounded facet (FBF) search. Suppose that there is a flux point $x = A$ at which all constraints coincide. Then, it satisfies the matrix equation $C \cdot A = V$. As discussed in Section 6.1 of Chapter 5, this can be solved by employing the Moore–Penrose pseudoinverse matrix C^+ as follows:

$$A = C^+ \cdot V \quad \leftrightarrow \quad C \cdot A = V \tag{7.4}$$

where the first equation is used to construct a putative solution, which only exists if the second equation is satisfied.

If such an apex point A exists, any point in the SS polytope can be reached from it by adding a multiple of a ray vector, and so by definition A is itself the SSK. In that case it makes no sense to describe the size or shape of the SSK. However, the cone still has a cross-sectional shape, and to characterise this we relax the compactness goal for this case. The inherent flexibility in capping hyperplane choice is exploited to extend the SSK beyond its minimal realisation as a single point, by truncating the cone at an arbitrary but small distance away from the apex. That is the rationale for step 3 in the tangent capping procedure.

A second noteworthy aspect is that additional bounded facets formed by capping hyperplanes play no role in fixing the capping radii. Testing in step 5 only involves the *real* bounded facets determined by the model constraints, not the *artificial* ones that merely form partitioning boundaries separating the SSK from the unbounded part of the SS. This is part of ensuring that the SSK (as far as possible) only reflects the biochemically meaningful implications of the model constraints.

A third point is that there are two recentering actions. The first in step 9 is needed because the SS coordinate origin may well fall outside of the region partitioned off by capping. Moving it back inside facilitates the elimination of redundant constraints in step 11. Once the final, fully closed specification of the SSK has been achieved, centering is further improved in step 12 to define an origin well away from polytope boundaries as the centre of the maximal inscribed hypersphere.

Occasionally, it happens that the maximal radius turns out to be negligible for a particular model. As pointed out in Chapter 3, Section 2.3, in this case inscribed hypersphere centering may fail to move the centre off the boundaries and so step 13 implements a heuristic to compensate for that. This heuristic makes use of the method for calculating the offcut radius along a given direction, as described at the beginning of section 2.4 of Chapter 3 in connection with centre refining.

Note that the centering performed as part of tangent capping is still not final; the final refinement can only be done after the chord determination discussed in section 7.2.

A final remark is that it may also happen that all capping radii turn out to be negligible. With inclusion of the ray space basis and constraint vectors, the capping rays form a complete basis for the flux dimensions in which the SS is embedded. Hence, in this case the SSK itself is so small as to be essentially just a single point in that space. A reasonable interpretation for what 'negligible' means in this context is that the variation of flux values away from the origin is similar to the numerical tolerance within which a flux value is considered constant, as discussed in Chapter 2. When this occurs, capping radii are assigned the default value as in the case of a true single point SSK in step 3. Then the SSK becomes an approximate simple cone, albeit with a small apical hypervolume instead of a single point apex.

7.2 Orthogonal Chords and Shape Characterisation

7.2.1 *Chords and Aspect Ratios*

Having reduced the solution space to a bounded kernel, the question arises: what is the shape of this multidimensional polytope? The concept of geometrical shape, familiar in 2D and 3D, seems much harder to grasp in higher dimensions. Even for four dimensions (4D), discussions of shape usually rely on analogies with lower dimensions and are not readily generalised further. Still, although they cannot be visualised, some intuitive understanding of the differences in shape between, for example, a hypersphere, hypercube, simplex and hyperellipsoid in any number of dimensions lies within reach.

One useful tool to discuss this is a set of mutually orthogonal, maximal length chords. For a hypersphere in N dimensions, all maximal chords have identical lengths, so *any* set of N orthogonal directions gives the same identical values. A hyperellipsoid has *one particular* set of such directions giving N distinct maximal chord lengths. A range of intermediate shapes in which maximal orthogonal chord lengths appear in distinct groupings of identical values are also possible. Similarly, for families of corresponding polytopes such as hypercubes or simplexes, some consistent patterns of how such maximal chord lengths are distributed can be found, as further discussed below.

For a general, irregularly shaped polytope in N dimensions, 2D and 3D analogues are not very helpful to describe it. Nevertheless, one can envisage choosing a set of mutually orthogonal axes, calculating the maximal length chord that spans the polytope along each axis direction, and rotating the set of axes until the set of chord lengths is maximised in some sense (e.g., the sum of lengths).

This set of chord lengths would convey a sense of the *size* of the polytope, namely the hypervolume it encloses. Moreover, some appreciation of its shape can be obtained by mutual comparison of the chord lengths. In particular, any pair of axes defines a 2D plane in the N-dimensional space. If all points in the polytope are projected onto this plane, this creates a 2D **shadow** of the polytope. The two projected chords retain their lengths but only give lower limits for the chord lengths of the shadow along their directions, as illustrated in Figure 7.1. The ratio of the pair of chord lengths defines an **aspect ratio** of the polytope, as seen along the direction of the vector in ND that is normal to the plane of the pair of axes.

For simplicity, one can arrange the set of axes in the order of decreasing chord lengths and take the aspect ratio of each pair in this descending order. Then, no aspect ratio will be less than 1, and a numerically large value indicates that the polytope is elongated in the first direction relative to the second.

Clearly, a plot of the chord lengths in descending order can convey at least a general impression of whether the polytope is approaching a regular shape such as a hypercube or even hypersphere (many nearly equal chords) or at the other extreme is very elongated or very flat along some directions.

The maximal aspect ratio will, by construction, be the ratio between the first and last chord lengths. Suppose that has a value of 100. Inspection of the entire sequence may reveal that the large aspect ratio is due to the first length being much larger than most of the others, in which case the polytope can be described as very elongated in that direction (like a prolate ellipsoid). At the other extreme, it may reveal that it is the last chord length that is much smaller than most of the others, in which case the polytope is very thin along this direction (like an oblate ellipsoid).

The latter case is particularly relevant for optimising the SSK. If there is a thin direction, it means that compared to all other directions in the reduced flux space, there is very little variation of the flux along this direction, over the extent of the SSK. The situation of flux components that remain fixed, as dealt with in Chapter 2, can be seen as just the extreme case of this where the SS occupies just a mathematical sub-hyperplane. Such cases were seen to be very common, and allow quite major reductions in SS dimensions by projecting the SS to such hyperplanes. So, it comes as no surprise that cases where the SSK has thin directions are also quite common.

In fact, it was remarked in Chapter 2 that the algorithms employed there may occasionally fail to detect a fixed flux in the case of a metabolic network with reversible reactions. If so, such a fixed flux component will be picked up by the analysis of maximal chord lengths yielding a near zero length.

The detection of thin directions is also in another sense a generalisation of the cases dealt with in Chapter 2. That analysis only involved single flux components, that is, a flux through a particular reaction of the metabolic model. Geometrically that represents a flux hyperplane orthogonal to one of the coordinate axes in the original, full flux space. By contrast, here the thin direction detected can have any orientation in flux space, not necessarily along one of the original axes in the full flux space.

This general orientation corresponds to a combination of flux components, that is, in biochemical terms indicates a biochemical pathway that becomes limited to an essentially fixed flux as a result of the combined constraints of the model. Conceivably, each of the fluxes that make up this pathway may still have a finite range of variation due to their participation in other pathways. In such a case, the fixed flux in this pathway would not be picked up in an alternative approach such as flux variability analysis (FVA).

It follows that for consistency, any sufficiently thin direction detected by chord length analysis should be treated as was done with *fixtol* (the fixed value tolerance first introduced in Chapter 1 in the discussion of table 1.1). In other words, the SSK should be projected onto a hyperplane orthogonal to the thin direction, with consequent further dimension reduction. The implementation of the SSK reduction discussed in section 1.5 allows for the software user to decide interactively whether to carry out such further reduction.

7.2.2 Flattening Thin Directions

The first problem is to decide when the SSK is sufficiently thin to be flattened out into a single hyperplane. A lower limit for this is clearly the numerical tolerance *fixtol* chosen for determining fixed fluxes. But if the remaining chord lengths are sufficiently large, it may be justified to flatten out even larger variations on the grounds of a large aspect ratio.

For perspective, note that the aspect ratio of the lengthwise, edge-on side view of a credit card is typically about 100. It seems plausible that this can, to a good approximation, be flattened out into a plane. However, it should be borne in mind that the flattening needs to be done by downcasting the constraint hyperplanes according to Eq. (2.9). This is not quite the same as constructing the polytope shadow. Instead, it amounts to taking a cross-section through a chosen point in the thickness of the SSK.

Figure 7.1 illustrates some of the concepts involved in flattening.

ABCD represents a 2D SSK, that is, a range of feasible fluxes, which is flattened to a one-dimensional (1D) range AE, by contracting the small range of variation d along the supposedly thin direction BF to a single point. During flattening, constraint lines AB and CD, respectively,

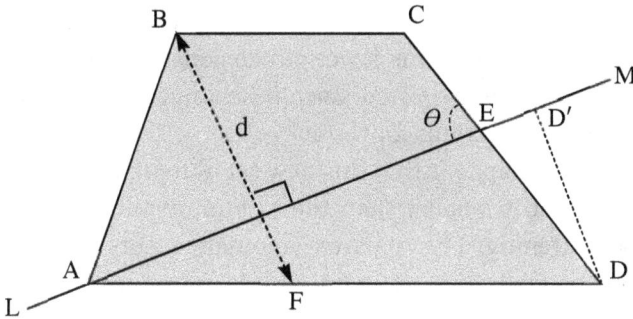

Figure 7.1 Flattening of trapezium ABCD to a line section AE. BF and AE are chords of maximal length, given a fixed choice of their mutually orthogonal directions. For legibility, the length d of chord BF is exaggerated; flattening is only applied if d is of the order of 1% of AE. The 1D shadow of the 2D polytope ABCD is the line section AD′. AE is only an approximate chord of the shadow, the approximation improving as d reduces.

become constraint points A and E on the line LM that is perpendicular to the thin direction BF. Constraint lines AD and BC become redundant and are eliminated. So, the original SSK specified by four constraints in two dimensions has been reduced to two constraints in one dimension.

With small d, all flux values on the line BF are slightly different, but coalesce to a single flux value during flattening. The approximation errors for all these points are less than d, which may be acceptable if d is small compared to the norm of the flux vectors involved. However, notice that for point D, which projects to a point D′ that is outside the projected SSK, the closest representative point is E, and this involves an additional error that for a small angle Θ may be much larger (though still proportional to d).

This example illustrates that the flattening approximation may need to be applied with caution, as depending on the details of a metabolic model, it may induce larger errors than suggested by a large aspect ratio.

In the bigger picture, though, what counts is not how close the geometries of the SSK are before and after flattening, but rather the extent to which the flattened SSK can still correctly generate the full SS by the addition of ray vectors. As the flattened SSK is by construction a subset of the unflattened one, all fluxes that it generates will remain members of the SS, that is, valid feasible fluxes. However, there may be a margin of SS points that can no longer be reached from the flattened SSK. As will be shown in the next chapter, for any given SS point, a deconstruction into its kernel and ray contributions can be found. This process will be used to estimate the size of this unreachable margin, and establish a criterion for when flattening is justified in any particular model.

When flattening is justified, the coordinate origin needs to be shifted to the interior of the resulting lower dimensional SSK and the determination of main chords repeated since its changed shape may produce different optimal directions for those.

It occasionally happens that this new set of main chord lengths contains a value that is smaller than the minimum value that was set to survive the flattening. This apparent anomaly occurs because the progressive orthogonalisation of main chords introduced in section 7.3, restricts the spatial directions that are available to find the next chord. This is a cumulative effect that particularly constrains the shortest chords so that some thin directions may be missed.

The remedy is that flattening is repeated automatically in this case, if necessary several times until no chords shorter than the chosen minimum remains. In this process, the maximal inscribed hypersphere diameter is also a useful guideline; when this is much smaller than the smallest detected chord length, it implies that some thin directions have been missed so far.

In actual metabolic model examples, the error estimates made by deconstruction shows that flattening errors tend to accumulate as more directions are flattened, and this often limits the number of dimensions that can be removed in this way, to a single-digit value. This may still leave a large maximal aspect ratio and other flattening indications unsatisfied, such as a discrepancy between the smallest chord length and the inscribed hypersphere diameter.

A pragmatic solution to this dilemma is to repeat the entire SSK calculation starting with Hop, Skip and Jump, with a fixed value tolerance, *fixtol*, increased to a value that is chosen to avoid excessive aspect ratios from the start.

This is an extension of the role originally intended for the *fixtol* variable, which was merely to take numerical inaccuracies into account. However, it can be justified in several ways. It is computationally much more efficient to eliminate thin directions at the solution space reduction stage, instead of carrying them through the subsequent stages such as BFBF search that are sensitive to the dimension count, only to be eliminated by flattening afterwards. Also, it limits the need for flattening to only thin directions that were not easily detectable at the reduction stage and so reduces accumulation of flattening errors.

An additional factor comes into play for large models with several thousand flux variables. These models can present convergence problems for the various LP calculations that are involved. This is linked to the fixed value tolerance, in that the LP calculations used to test for a fixed value need to converge to an accuracy at least an order of magnitude better than *fixtol* in order to make the test viable. So, an increase in *fixtol* is sometimes crucial to allow LP convergence in the Hop, Skip and Jump algorithm and may need to be introduced purely for that reason in large models.

Except for that situation, the general strategy is to first complete an SSK calculation with the default *fixtol* value in order to obtain a full set of main chord lengths. When inspection of these and the resulting aspect ratios show an excessive size range, *fixtol* is set to a value typically between 0.1% and 1% of the maximal chord length. Repeating the calculation, the maximal aspect ratio is hence guaranteed to be below

1000. Further fine-tuning can then be carried out by flattening any remaining thin directions, particularly any such that are clearly in a separate group from the main sequence of chord lengths.

7.3 Calculating Maximal and Main Chords of a Polytope

The problem of how to calculate a maximal chord for a polytope, given its halfspace representation, now needs to be addressed. An exact LP-based method briefly discussed below can be applied for single-digit dimensions. However, for larger dimensions, only more approximate methods are tractable and two such approaches are also presented below. The chords produced by these approximations are still exact chords, spanning between two points on the periphery of the polytope, but while large are no longer guaranteed to be the ones of maximal length. To emphasise the distinction, in the work below, the results of the approximate methods are designated as **main chords** while the term **maximal chords** is reserved for the exact calculation results.

The problem of finding a maximal chord is reminiscent of the problem to find a maximal jump from a given initial point for the Hop, Skip and Jump algorithm discussed in Chapter 2. However, for a chord, both endpoints are to be freely chosen.

This aspect is easily dealt with by doubling the space dimensions. Given the two chord endpoints x_1 and x_2 in an N-dimensional flux space, each of which satisfies the feasibility constraints $C \cdot x \leq V$ that define the SSK polytope, the chord is represented by the vector $y = (x_1 - x_2)$. In the compound $2N$-dimensional space obtained by joining the vector representations together, the chord is created by applying the $N \times 2N$ matrix multiplier Y represented in terms of the N-dimensional identity matrix I_N by:

$$y = x_1 - x_2 = \begin{pmatrix} 1 & 0 & \dots & -1 & 0 & \dots \\ 0 & 1 & \dots & 0 & -1 & \dots \\ \dots & \dots & \dots & \dots & \dots & \dots \end{pmatrix} \begin{pmatrix} x_1 \\ x_2 \end{pmatrix}$$

$$= \begin{pmatrix} I_N & -I_N \end{pmatrix} \cdot \begin{pmatrix} x_1 \\ x_2 \end{pmatrix} = Y \cdot \begin{pmatrix} x_1 \\ x_2 \end{pmatrix} \tag{7.5}$$

As discussed in Chapter 2, to efficiently formulate maximising the chord length as an LP problem, it is the sum of absolute values of y-components (i.e., the L1-norm) that is maximised rather than the Euclidean or L2-norm.

The discussion so far focused on finding a single maximal chord. The desired set of orthogonal chords is generated by adding a further constraint after each chord is found, to keep all subsequent chords orthogonal to it. That is described as a **progressive orthogonalisation** strategy.

This is not quite equivalent to simultaneously optimising the orientation of an orthogonal system of axes as envisaged earlier, since the maximisation of each subsequent chord length is more constrained than the one before. Progressive orthogonalization tends to produce a descending sequence of chord lengths. It may be seen as maximising a weighted sum of chord lengths (with a descending sequence of weights) rather than a simple sum. Nevertheless, it is the same approach used in the standard statistical method of principal component analysis, which addresses a somewhat similar problem of establishing mutually orthogonal directions along which a data set varies maximally.

For later use, the orthogonality constraint also needs to be formulated in the compound space. Let y_1 be the N-vector specifying the maximal chord obtained in the first chord calculation. Then, the next chord $y_2 = (x'_1 - x'_2)$ has to be orthogonal to it, that is, $y_1 \cdot y_2 = 0$, and this is equivalent to the $2N$-vector equation

$$\left(y_1 \quad -y_1 \right) \cdot \begin{pmatrix} x'_1 \\ x'_2 \end{pmatrix} = 0 \tag{7.6}$$

as is easily seen by comparing component expansions of Eq. (7.6) and $y_1 \cdot y_2$, using Eq. (7.5).

Note that since chord determination involves optimising coordinate differences, it is largely independent of the position of the coordinate origin. **Chords** do not necessarily intersect at the origin, or even at all. This distinguishes them from the special case of **diameters**, which do.

Once a set of main chords has been calculated by either of the methods described below, the vector basis in terms of which the SSK is specified is reoriented to coincide with the main chord directions. This is done to conform to the requirements of the D-specification discussed in Chapter 1 and leading to Eq. (1.12).

7.3.1 *Mixed Integer Linear Programming Determination of a Maximal Chord*

As discussed in Chapter 2, Section 4, LP *minimisation* of a sum of absolute values would be fairly straightforward by invoking an auxiliary variable. In the present case, this is an N-dimensional vector z defined by component-wise constraints $z \geq y$ and $z \geq -y$. This pair of constraint sets restricts z to the range $(|y|, \infty)$ so that minimising z gives the minimum $|y|$ value. This works because whether y is positive or negative, one of the pair of constraints amounts to $z \geq |y|$ and the other becomes redundant.

In order to *maximise* $|y|$, one could instead try to restrict z to the range $(-|y|, |y|)$ and then maximise z. However, that does not work because, of the four constraints that are needed to enforce the range, one pair becomes mutually exclusive whatever the sign of y is. The standard remedy for this is to invoke a further auxiliary variable, in this case a vector b with N binary components that serve to eliminate the contradictory constraints. Explicitly, the constraints are given by

$$
\begin{aligned}
z \geq y &\quad \Rightarrow \quad Y \cdot (x_1 - x_2) - z \leq 0 \\
z \geq -y &\quad \Rightarrow \quad -Y \cdot (x_1 - x_2) - z \leq 0 \\
z \leq y + M * b &\quad \Rightarrow \quad -Y \cdot (x_1 - x_2) + z - M * b \leq 0 \\
z \leq -y + M * (1-b) &\quad \Rightarrow \quad Y \cdot (x_1 - x_2) + z + M * b \leq M
\end{aligned} \tag{7.7}
$$

Here, M is a scalar value chosen a priori to be larger than twice the maximal value of $|y|$.

Taking the case $N = 1$ for simplicity, one can trace through these equations separately for the cases $y > 0$ and $y < 0$. One finds that the first case requires that $b = 0$ to avoid mutually exclusive equations and the second that $b = 1$. In each case, only one equation survives redundancy and ensures that $z = |y|$, so that maximising z delivers the required Max($|y|$).

Combining Eq. (7.7) with the polytope constraints, maximising the chord length requires solving the LP problem of maximising an objective vector of the form O = {0,0, ..., 0,0, ..., 1,1, ..., 0,0, ...}, where the non-zero entries select out the z-components in a $4N$-dimensional space (x_1, x_2, z, b), subject to the extended constraint matrix:

$$
\begin{pmatrix}
C & 0 & 0 & 0 \\
0 & C & 0 & 0 \\
I_N & -I_N & -I_N & 0 \\
-I_N & I_N & -I_N & 0 \\
-I_N & I_N & I_N & -M*I_N \\
I_N & -I_N & I_N & M*I_N
\end{pmatrix}
\cdot
\begin{pmatrix}
x_1 \\
x_2 \\
z \\
b
\end{pmatrix}
\leq
\begin{pmatrix}
V \\
V \\
0 \\
0 \\
0 \\
M
\end{pmatrix}
\qquad (7.8)
$$

One complication is the need to estimate a value for M, but as only a rough estimate suffices this is not problematic. One can, for example, use trigonometry to find the radius to the intersection of each pair of constraint hyperplanes and use a multiple of the maximum of those. More significant is that this LP is in a $4N$-dimensional space with a correspondingly large increase in constraint count compared to the underlying polytope specification in N dimensions.

But the crucial stumbling block is that the vector b has to be restricted to have binary components 0 or 1 only, otherwise the required reductions of Eq. (7.7) do not take place. This means that the problem becomes a Mixed Integer Linear Programming (MILP) problem, which is far more difficult computationally.

One standard approach to facilitate the solution of MILP problems is to relax the integer requirement to allow real values (in this case for the b-vector). This would normally produce an upper limit for a maximisation problem, and it becomes a matter of establishing how much higher the result is than the actually desired MILP maximum.

However, for the maximal chord problem it turns out that the relaxed maximum of z is a constant, the value $0.5\,M$, independent of the value of y, and so this does not give information about the maximum of $|y|$ any more.

It follows that for maximal chords it remains necessary to solve the full MILP problem. Even in 2D there are examples where this is problematic, and it turns out that beyond single-digit dimensions the MILP approach often becomes quite intractable.

7.3.2 Polytope Flipping — an Approximate LP Chord Calculation Method

In order to overcome the tractability hurdle, consider a rather drastic simplification of the problem: simply maximise the sum of chord components y_i, rather than their absolute values. This would eliminate the need for auxiliary variables, halving the dimensions and more importantly restoring it to a simple LP maximisation problem.

The obvious snag is that it is conceivable that the chord of maximal length may have positive and negative components that may cancel to yield a zero sum, indistinguishable from, for example, the trivial chord of zero length that connects two neighbouring periphery points.

Note that the discussion may be restricted without loss of generality to positive component sums, as any chord with a negative sum can simply be reversed by swapping its endpoints to give an equivalent positive chord length.

To discuss this further, it is useful to classify the infinite number of chords into a finite number of groups, according to the 'hyperquadrant' of the 2N-dimensional coordinate space in which they fall. The term **hyperquadrant** refers to a subset of the coordinate space, consisting of all vectors that share the same sign for each corresponding vector component. Identify a particular hyperquadrant by a representative vector q of a form such as $\{1,-1,-1,\ldots, 1,1\}$. Specifically take $q_i = 1$ for the vectors with components $y_i \geq 0$, and $q_i = -1$ for $y_i < 0$. Vector q is termed the **quadrant vector** below.

For all chords that fall in the all-positive hyperquadrant represented by $Q = \{1,1,1, \ldots\ldots, 1,1,1\}$, maximisation of the component sum is equivalent to maximising the sum of absolute values. So, the proposed simplification delivers the correct maximal length chord from this subset of chords (as well as the equivalent set belonging to the all-negative hyperquadrant).

For other hyperquadrants, only a lower limit to the maximal length is obtained. The discrepancy may be anticipated to be worst in hyperquadrants where the numbers of positive and negative q_i are nearly balanced, enhancing the opportunities of cancellation during summation.

However, the orientation of the polytope relative to the coordinate axes is immaterial for chord determination. So, it is in principle possible to rotate the polytope so that chords originally belonging to any other hyperquadrant now occupy the all-positive quadrant, and then repeat the LP maximisation. Doing this systematically for all hyperquadrants and taking the maximum chord length over all hyperquadrant maxima will give the correct maximal chord length for the polytope.

The discrete rotation in N dimensions as required to rotate an arbitrary chosen quadrant vector q to the all-positive vector Q, is represented by a rotation matrix R for which an explicit form is supplied by suitable mathematical software systems such as *Mathematica*. By construction it satisfies

$$q \cdot R = Q \tag{7.9}$$

When applied to any vector x in the chosen hyperquadrant, the effect of R is merely to change the arithmetic signs of components similarly as for q in Eq. (7.9). Hence, the entire hyperquadrant is rotated to coincide with the all-positive hyperquadrant by this rotation. In terms of the polytope specification, since each row of C defines the orientation of a constraint hyperplane, the entire polytope is rotated by the matrix rotation $C \cdot R$. For brevity, such a discrete rotation is henceforth called a **flip** of the polytope.

Returning to the simplified calculation of each maximal chord, it consists of a series of LP problems in which the objective vector in the composite or y-space is given by the $2N$-dimensional vector {1,1,...1, –1, –1,... –1} that consists of N positive unit entries followed by N negative ones.

The constraint equations for the sequence of chord determinations then have the form:

$$\begin{pmatrix} C \cdot R_n & 0 \\ 0 & C \cdot R_n \\ K_k & -K_k \end{pmatrix} \cdot \begin{pmatrix} x_1 \\ x_2 \end{pmatrix} = \begin{pmatrix} V \\ V \\ 0 \end{pmatrix} \tag{7.10}$$

Equation (7.10) incorporates the progressive orthogonalisation of maximal chords described before. In particular, it specifies the constraint set for the $(k + 1)$th round of determining maximal chords. Here, K_k represents a $k \times N$ matrix in which each row is occupied by a previously found chord vector y_k. For the first round, $k = 0$, and matrix K_0 is by definition empty, so its row is omitted from Eq. (7.10). For subsequent chords, this row expresses the orthogonality condition as given in Eq. (7.6).

In each round, R_n signifies the rotation matrix for the nth flip that is executed. The start of the round has $n = 0$ for which there is no rotation, so $R_0 = I_N$, the N-dimensional identity matrix.

An exhaustive determination of the maximal chord in each round would require R_n to cycle through all hyperquadrants. But as the number of hyperquadrants increases exponentially (as 2^N) with the dimension count, this would soon become prohibitive. Fortunately, it turns out that the maximal length collected over a sequence of flip steps converges surprisingly quickly.

Two explanations for this may be advanced. First, recall that in each orientation all hyperquadrants are in effect sampled by the LP maximisation, only the sampling is biased in favour of those with quadrant vectors of predominantly identical arithmetic signs. When flipping the polytope so that a particular hyperquadrant occupies the prime position, the biasing of all other hyperquadrants is also changed so improving the chances of detecting a previously missed long chord. A second factor is that for each subsequent chord the search space becomes more constrained by the accruing orthogonality constraints, which also enhances the probability of detecting a previously missed long chord. Supporting evidence for this happening is that it is observed that the sequence of chord lengths generated by the procedure, occasionally shows small deviations from a monotonous decrease.

The observed convergence suggests that the calculation can be made more tractable for large dimensions by limiting the number of flips that are executed. The actual rotations are selected by random sampling. Each rotation is uniquely specified by a particular quadrant vector q that is rotated to coincide with the direction of the all-positive quadrant vector Q. Hence, each selection amounts to choosing q in an appropriately random way.

To make this choice, the first step is to randomly choose a subset of approximately half of the vector components of Q, to change from +1 to −1. This is because the resulting vector q belongs to a hyperquadrant that is likely to be most severely affected by the bias introduced by component sum maximisation. So, choosing half the components to undergo a sign change is designed to maximally counteract the inherent bias of the current polytope orientation.

For low values of N, this will limit the set of polytope orientations that are sampled to quite a small set that would be repeatedly chosen if the number of flips that are executed is larger than this set. In order to sample the orientations from a larger set while maintaining the emphasis on bias reduction, the second step is to add to the random quadrant vector of the first step, another N-vector of which each component is a random number in the range (0, 0.1), that is, a 10% random perturbation of each component.

Repeating this process for each flip generates a sequence of discrete but quite distinct polytope orientations, each differently biased, so that the chord length that is maximal over all of them is quite likely to be the correct maximal chord.

It is at this point that the procedure becomes an approximation. It always delivers a valid chord length, but without having made a completely unbiased exploration of all hyperquadrants, the final result can only be regarded as a lower limit on the maximal chord length. Experience has shown that this lower limit often coincides exactly, and else approaches closely, the actual maximal chord length that is obtained from the MILP method. To emphasise the fact that full maximisation cannot be guaranteed, the chords obtained from the polytope flip method are called the **main** polytope chords.

For a polytope in N dimensions, there are N main chords to be calculated and for each of these, multiple LP calculations in $2N$ dimensions are required, one for each flip. Even though the SSK dimensions have usually been reduced to a double-digit number, which is quite small by the standards of LP implementations, the total computing effort does mount up. So the number of flips needs to be chosen carefully.

Anecdotally even as few as three flips deliver good results, judged both by comparison with MILP results for cases up to $N = 10$, and regarding consistency when making many repeated runs for dimensions in the range $50 \leq N \leq 60$.

The proliferation of the number of hyperquadrants with increasing dimensions suggests that more flips are appropriate as N increases. Another consideration is that in progressive orthogonalisation, accurate determination of the chords found first is more important than for later ones, because each chord influences all subsequent ones through the mutual orthogonality constraints. This should allow the number of flips to decrease as the counter k in Eq. (7.10) is incremented.

A pragmatic heuristic to accommodate these considerations is to firstly allow the user to choose an absolute maximum number, *flipmax*, for which a reasonable default value is *flipmax* = 50. The actual number of flips for the first chord is taken as Minimum[*flipmax*, 2*N], and this number is linearly decremented for each subsequent chord to reach a value 3 for the last one. In addition, the flipping loop is exited early if the same maximum chord length is found repeatedly. The critical number of repeats for this decision is chosen as at least 3, but is increased logarithmically with N.

The description above supplied the rationale for the polytope flipping method to calculate main polytope chords, but may not in itself be convincing that it will work in practice. Some concrete examples are presented in section 7.3.3 to address this.

7.3.3 *Example Main Chord Calculations*

To test the proposed calculation, it is first applied to some examples where values or patterns can be established independently, from geometrical arguments. A first, somewhat trivial confidence-building case is a set of regular polygons in 2D, represented by increasing numbers of constraint vectors in $C \cdot x = V$ set at equal angle intervals and with all entries in V equal to 1. Starting from 3 sides giving an equilateral triangle, this produces a square, pentagon,hexagon, etc.

Geometry shows that where the vertex count is a multiple of 4, the two maximal chords are equal; and since polygons with increasing vertex numbers approach a circle of radius 1, the two chords have different lengths in other cases (but with decreasing differences) and their lengths converge to the value 2. The LP main chord calculation results are shown in Figure 7.2 and follow exactly this pattern.

More challenging, and more relevant to SSK characterisation, is to extend this to higher dimensional examples. Consider the case of hypercubes, bounded by pairs of hyperplanes orthogonal to each coordinate axis and displaced 1 unit from the origin in either direction. The vertex positions for such a hypercube are given exactly by the quadrant vectors described in section 7.3.2. The diagonals of such a hypercube are the vectors that connect pairs of vertices with components of uniformly opposing signs; the main diagonal is the vector 2{1, 1, ..., 1} with length

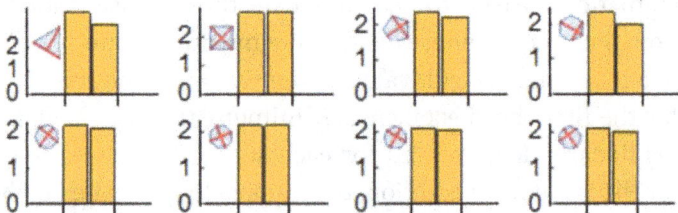

Figure 7.2 Bar charts of calculated main chord lengths for regular 2D polygons with, respectively, 3,4,5...10 vertices, in normal reading order. Each bar chart has its corresponding polygon as an inset, with the calculated chord pair shown in red.

$2\sqrt{N}$ and is the longest chord of this hypercube. All diagonals have the same length, which is that of the maximal chord, but may or may not be mutually orthogonal.

Starting with the 2D case (a square), its diagonals are the vectors {1, 1} and {1, –1}, which are mutually orthogonal. So, in this case, there are two equal longest chord lengths in the orthogonal set. In 3D, the diagonals are {1, 1, 1}, {–1, 1, 1}, {1, –1, 1} and {1, 1, –1}. As the minus entries in the co-diagonals are not in balance with the positive entries, none of them are orthogonal to the main diagonal. Hence, there is a single longest chord length and all other maximal orthogonal chords have shorter lengths. In 4D, it is again possible to find three co-diagonals that are orthogonal to the main diagonal as well as each other, namely {1, –1, 1, –1}, {1, –1, –1, 1} and {1, 1, –1, –1} and so the 4D hypercube has four equal longest orthogonal chords.

Generalising to higher dimensions, in all cases the main diagonal can be chosen as the chord of longest length overall. Whether there are other co-diagonals orthogonal to it, depends on whether minuses can be distributed over the components in such a way that full cancellation takes place when taking the vector dot products.

Clearly, with an odd number of dimensions this is not possible so there is just the single maximal chord length.

For an even number of dimensions, the co-diagonal obtained by changing the sign of every second component is always orthogonal to the main diagonal, so there are always at least two equal maximal orthogonal chord lengths.

As shown above, for $N = 4$, all the maximal orthogonal chords in fact have equal lengths, and that turns out to be true for N any multiple of 4.

It is not true, however, of even N values in general; for example, it is easily established by writing out the components that for $N = 6$ there are only the two equal maximal lengths arising from the main diagonal and its only orthogonal co-diagonal.

These remarks do not cover all the regular patterns that can be discovered, but do give a yardstick against which the actual maximal orthogonal chord lengths calculated by polytope flipping can be measured for higher dimensions.

The result is shown in Figure 7.3. It is seen that all the calculated main chords follow exactly the patterns just described. Moreover, the actual longest chord lengths found are in exact agreement with the geometric value $2\sqrt{N}$.

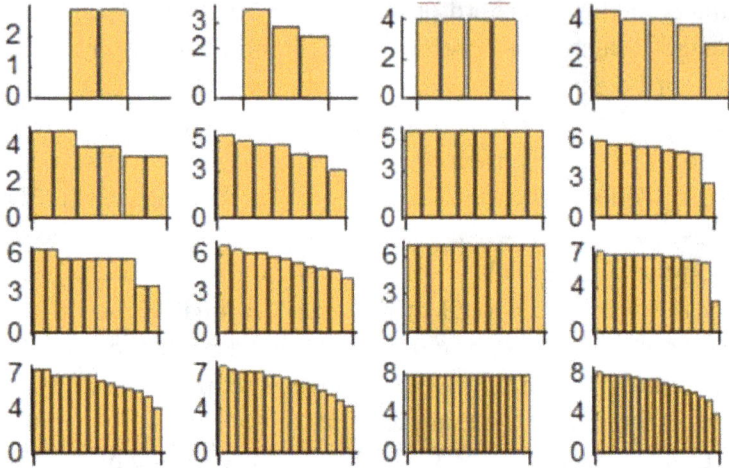

Figure 7.3 Bar charts of main chord lengths calculated by polytope flipping for hypercubes in, respectively, 2,3,4,5...17 dimensions, in normal reading order. The columns of figures alternate between even and odd dimension counts; the third column shows dimensions that are multiples of 4.

A second test case is obtained by considering simplexes in multiple dimensions. These are constructed by choosing $N + 1$ constraint vectors as unit vectors radiating at equal angles from the coordinate centre, and with constraint hyperplanes set at unit perpendicular displacement as specified by setting all entries in the values vector to 1.

For $N = 2$, this is an equilateral triangle. As illustrated by the first case shown in Figure 7.2, the first maximal chord is a side of the triangle and the second (the maximal chord that is perpendicular to it) is the (shorter) height of the triangle.

For $N = 3$, the simplex is a tetrahedron. Chord 1 is any edge of the tetrahedron that connects two vertices; as a tetrahedron has four vertices, there is a second edge that involves the other two vertices and it happens to be perpendicular to the first. So here chords 1 and 2 are equal, but the third is shorter.

Similarly, in N dimensions, there are $N + 1$ vertices; the edges involving the $n = $ Floor$[(N + 1)/2]$ non-overlapping pairs of vertices are mutually orthogonal and give n equal length maximal chords, followed by a progression of smaller ones. The actual maximal length can be calculated from the fact that in N dimensions, an edge spans an angle of Cos$[\theta] = -1/N$ at the centre of the simplex, and by construction the perpendicular radius from the centre to any side is 1. This leads to an edge length $\sqrt{N(N + 1)}$.

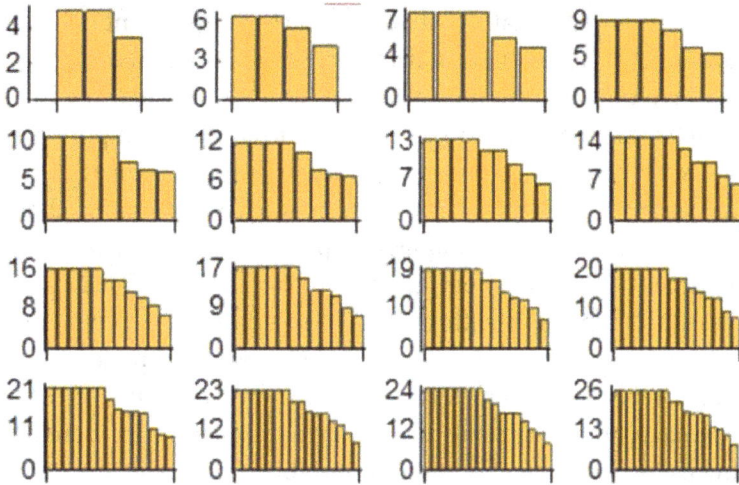

Figure 7.4 Bar charts of main chord lengths calculated by polytope flipping for simplexes in, respectively, 3,4,5…18 dimensions, in normal reading order. The columns of figures alternate between odd and even dimension counts.

Hence, for the family of simplexes, the expected pattern is that in even dimension counts half of all maximal orthogonal chord lengths are equal to the edge length, while the rest have declining values. For odd dimension counts, the number of equal maximal lengths is the same as for the next highest even dimension.

The actual main chord calculation results are shown in Figure 7.4. There is again exact agreement between the longest lengths found and the edge length formula. By and large the number of duplicate longest lengths also follow the predicted pattern, but there are small deviations. For example, in the case $N = 9$, only four longest equal lengths are found while the fifth is slightly shorter and equal to the next longest value. The cases $N = 11$ and 16 show similar behaviour. These deviations are for a particular run of the main chords calculation; if repeated the deviations may occur for different dimension counts. This is clearly a result of the random choices made in the polytope flipping procedure. Trials have been made with larger dimension counts up to the range 80–100, and no significant increase in the severity of the deviations were found – still only a single member of the longest length chord set was occasionally missed and replaced by a chord of slightly shorter length.

During the trials, the actual number of flips executed was monitored and the largest number required for any chord in N dimensions was typically in the range between N and $1.5N$, for either of the two polytope families. This is very small compared to the hyperquadrant count of 2^N, and presumably the minor deviations pointed out above could be reduced or eliminated by increasing the number of flips that are executed. However, the price to pay for this in terms of computing effort does not seem to justify the small performance improvement.

The examples shown above mainly test the sets of longest chords as predicted from geometry. To judge the performance of the main chord calculation on the subsequent, shorter, orthogonal maximal chords in each case, a comparison with the MILP results for the hypercube and simplex families can be used. The MILP computing time for a simplex turns out to be several times larger than for a hypercube of the same dimension, and the time tends to roughly double for a unit increment in the dimension count. Due to this unfavourable scaling, the MILP method becomes untenable beyond about $N = 20$, since the time to find chords then already exceeds the entire SSK calculation by a factor of three or more. By contrast, the main chord calculation consumes negligible time.

Two typical examples at the respective extremes of this range are shown in Figure 7.5. In the case of the hypercube, the two sets of results are very similar. It might be expected that the MILP result should always be the largest because it is only limited by the accuracy of the underlying LP, while the main chord calculation involves additional approximations. However, this is clearly not the case; there are small discrepancies in both directions.

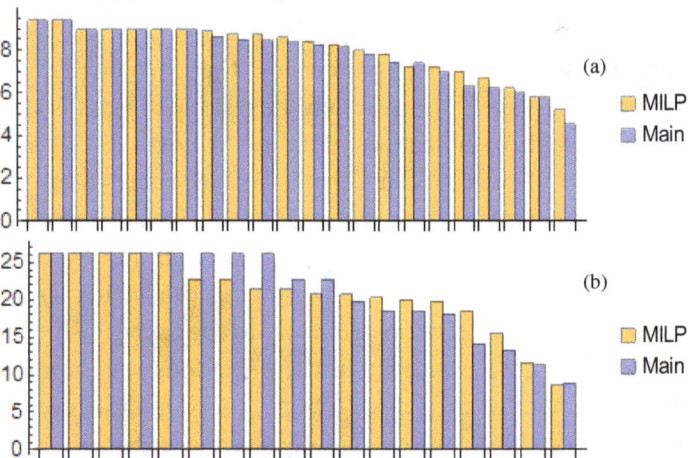

Figure 7.5   Bar chart comparison of MILP and polytope flip main chord length calculations for regular polytopes: (a) A hypercube in $N = 22$ dimensions; (b) A simplex in 18 dimensions.

One explanation is that even where the chord lengths agree, chord directions in hyperspace may differ. In such a case, the orthogonality requirement put on subsequent chords is different and can produce somewhat different results.

For the simplex, the differences are much larger. According to the geometrical argument presented earlier, there should be nine equal longest lengths in 18 dimensions. The main chord calculation finds only eight, but MILP does worse with five.

As the progressive orthogonalisation can cause individual lengths to vary, the mean chord length may be a better basis of comparison. In the cases shown, the mean main chord length is 98% of the MILP value for the $N = 22$ hypercube, and 100.7% of the MILP value for the $N = 18$ simplex. A large number of trials in fewer dimensions shows similar discrepancies between MILP and polytope flip main chord calculations of ±1–2%, with no discernible pattern.

Regarding computing times, for the MILP calculations of Figure 7.5, these were 242 and 228 sec, respectively, while the main chord calculations took 1.2 and 0.4 sec, respectively. Even for low dimensions, the approximate method is faster by an order of magnitude, and as its time requirement grows roughly linearly, compared to the exponential growth for MILP, the gap widens correspondingly at higher dimensions.

The question arises whether the good performance demonstrated so far might only hold for highly regular polytopes. Trials with irregular 2D polygons, in fact show exact agreement between MILP and main chords in all cases tried.

To extend the test to cases directly relevant to metabolic models, a couple of models (discussed in more detail in a subsequent chapter) where the SSK dimensions happen to reduce below 10 have also been subjected to both MILP and polytope flipping calculations.

Figure 7.6 shows the results, and by and large there is excellent agreement.

The most obvious discrepancy is seen in Figure 7.6(b) where the MILP calculation shows four zero-length chords, whereas the main chord values vary from 0.008 to 0.025. These values imply aspect ratios of larger than 500 rather than the infinite values MILP imply, but still ample to induce a need for flattening. Figure 7.6(c) shows that after flattening good agreement is obtained.

Closer inspection shows that in all three cases, there are still minor differences.

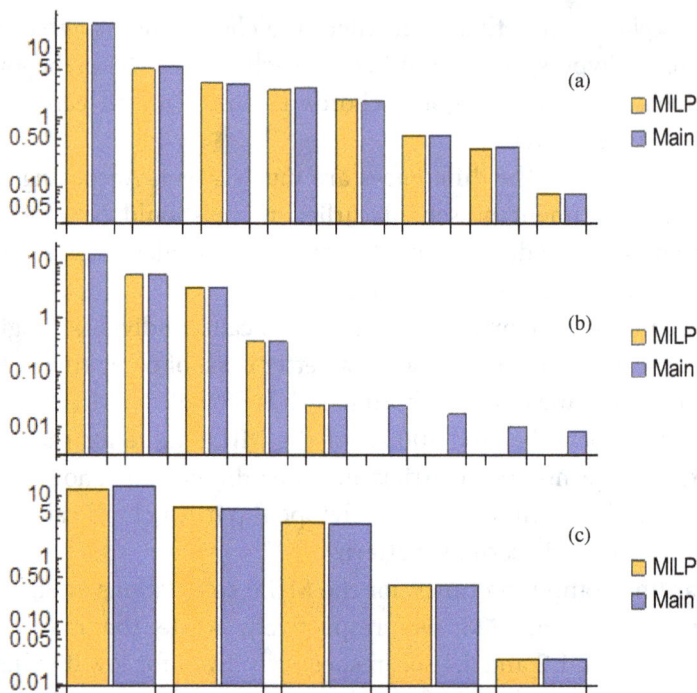

Figure 7.6 Logarithmic bar chart comparison of MILP and polytope flip main chord length calculations for various metabolic model SS kernels: (a) A red blood cell model, BIGG iAB_RBC_283; (b) Clostridium model BIGG iHN637, before flattening; (c) iHN637 after flattening.

The explanation advanced before about the effects of differing orthogonality constraints does not apply to the very first chord length, which represents an unconstrained overall maximum. Intriguingly, in both cases (a) and (c) in Figure 7.6, the main chord value is in fact larger than the MILP value. The respective values are 22.6 versus 23.2 in the first case and 12.7 versus 14.0 in the second, discrepancies too large to ascribe to LP tolerances. This looks anomalous, because MILP is meant to be exact while the main chords involve approximation.

However, actual testing confirms that the endpoints of all the main chords (for both methods) are indeed valid feasible points, so these particular chords were simply missed by MILP, as was the case also for Figure 7.5(b). Although relatively rare, this situation is observed in particular for high aspect ratio polytopes, such as typically result from metabolic models.

A plausible explanation is that the original orientation of the SSK polytope was unfavourable and produced inaccuracies with MILP, whereas the polytope flip method benefits from its sampling of multiple orientations. This is supported by the observation that if the complete SSK calculation is repeated (which due to the random aspects incorporated in that, gives a slightly different SSK specification), the MILP value for the Figure 7.6(c) case, for example, increases to 14.0 in complete agreement with the main chord value.

To summarise, the results of this section show that the polytope flip main chord approximation gives excellent results, in some cases even surpassing the supposedly exact MILP method. In addition, it is faster by one or more orders of magnitude. It scales quite well with dimension count, allowing application up to $N = 80$ or even beyond, whereas MILP becomes untenable around $N = 20$.

7.3.4 *Diameter-Based Main Chord Calculations*

For some SSK examples, the dimension count can run into three-digit numbers, and then even the main chord calculation becomes cumbersome.

An even quicker alternative is to exploit the fast calculation of diameters, avoiding LP by using simple geometry, as described in Chapter 3, Section 2.4. Recalling that a diameter is a special case of a chord, having the additional requirement that it passes through the coordinate origin, such diameter values can only be expected to give a lower limit to the length of the similarly oriented chord. As polytope shape characterisation relies on relative lengths along orthogonal directions rather than absolute values, this may nevertheless give an adequate approximation, provided that the origin is well-centered. In particular, it should be adequate to identify any thin directions that can be flattened out.

In fact, the centre refining procedure described in Chapter 3 provides a ready-made set of periphery points, in pairs that each defines a diameter, and constructed to cover the hyperangle space fairly uniformly. It is relatively straightforward to choose the largest such diameter as the first chord length, then project the remaining periphery points to the hyperplane orthogonal to it, and repeat until N orthogonal diameters have been found.

One problem with this simplistic idea is that projecting periphery points does not usually produce points on the perphery of the polytope. Instead, one can consider the diameter projections to furnish a sample set of directions only, orthogonal to the preceding subset of chords and recalculate the diameters along all of these directions after each projection.

However, there is a more fundamental logical flaw, namely that in fact centre refining and the chord calculation are intertwined, and it makes more sense to do the chord calculation first, as it is independent of the position of the origin. Moreover, the chord endpoints form an important input of periphery points that are kept fixed during centre refinement. The importance of chord endpoints is that they include polytope vertices that define the extent of the polytope in N mutually orthogonal directions. It is clear that the endpoints of the first chord will always be two vertices; all subsequent chords have at least one vertex as an endpoint, and where a chord lies along an edge such as for simplexes, both endpoints are again vertices. So, at least $(N + 1)$ and possibly more polytope vertices are supplied by the chord calculation and contribute to successful centering.

A pragmatic way to deal with this dilemma is to use the LP-based chord calculation for a pre-determined maximal number of chords, *chordmax*, first. When $N > chordmax$, the endpoints of the incomplete set of chords are used to perform a centre refinement. The full set of peripheral points that this delivers is used to calculate diameters that approximate the remaining $(N - chordmax)$ chords using the projection and direction sampling described previously.

A problem that can arise is that as each projection reduces the number of available sampling directions, these may run out before the full complement of N orthogonal chords has been reached. Two strategies are employed to alleviate this.

First, the set of constraint vector directions are combined with the peripheral point directions to compile the set of diameter sampling directions. Since the constraint vectors on their own already form a complete basis for the space in which the SSK is embedded, this guarantees an overcomplete set that is unlikely to run out.

Secondly, after each new chord determination, the centre refining is repeated with the incrementally enlarged set of chords. In effect, this samples different locations for the origin and reduces the restriction to diameters that all share a common intersection point.

In the rare case that unresolved directions still remain, these are simply determined as arbitrary directions orthogonal to all the known chords, and assigned the inscribed hypersphere diameter as chord lengths.

As a result, in the end even the diameter-based chords will not actually be diameters through the final refined origin. That is of no consequence as the same is true of the initially determined LP-based main chords.

In practice, it is found that the diameter-based chords are usually shorter than freely maximised chords of the same orientation, by a considerable single-digit numerical factor. As the range of chord lengths can range over several orders of magnitude for typical metabolic model SSKs, this is not as significant as it may seem. Nevertheless, the trade-off between accuracy and computing effort embodied in the choice of a *chordmax* value is best left as an interactive decision made by the user on a case-by-case basis.

To illustrate how LP-based and diameter-based chord lengths compare, Figure 7.7 shows the results for the Geobacter model that was used for demonstration in previous chapters. In all cases, the first (longest) several lengths were calculated by LP, with a changeover to the diameter-based method at, respectively, *chordmax* = 15 and 40 (out of the total of 56 chords shown).

In both cases, there is a sharp drop at the changeover point in the diameter-based length compared to the more accurate LP value. However, the overall trends are quite similar, while the gap between the methods reduces towards shorter lengths. A consequence of this is that decisions about flattening of insignificant dimensions can plausibly be based on the diameter-based values. In favourable cases, flattening may reduce the dimensions to such an extent that all remaining chords can be found by the LP method.

Figure 7.7 also shows that each of the two methods shows excellent internal consistency between independent runs, despite the random choices involved.

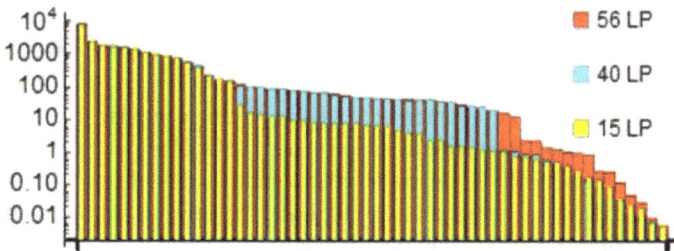

Figure 7.7   Logarithmic bar chart comparison of calculated chord lengths for the Geobacter model. Red bars show the case where all 56 chords were determined by the polytope flip LP method, whereas for the cyan bars the first 40 were found by LP and the rest are estimated from diameters, and for the yellow bars only the first 15 lengths are LP calculated.

7.4 Graphical Representation of the SSK Geometry

Several aspects of the SSK geometry can be displayed visually by graphical depiction of the calculated chords and diameters. The results for the red blood cell (RBC) model (BIGG iAB_RBC_283) are shown in Figure 7.8.

The most obvious feature is the number of bars, which corresponds to the SSK dimension count, that is, the value 8 in the example.

The first bar chart shows the calculated main chord lengths, displayed in different colours in cases where the diameter-based approximation was used for some. The flux space direction for each chord is not shown in the chart, but is accessible as the corresponding row vector of the matrix B in the D-specification of the SSK as given by Eq. (1.12).

Figure 7.8 (a) shows that regarded as a geometrical object, this SSK is quite anisotropic, with a large variation of main chord lengths. To quantify this, the geometric mean of aspect ratios formed by all pairs of main axes is also shown in Figure 7.8.

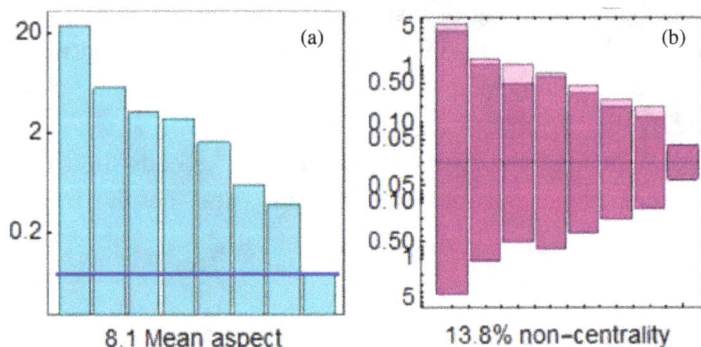

Figure 7.8 Graphical representation of SSK geometry, here applied to the red blood cell model. (a) The main chord lengths in descending order, along mutually orthogonal directions. Cyan bars depict LP-calculated chord lengths, while diameter-based chord lengths (if any) are shown in magenta. The blue line is the diameter of the independently calculated, maximal radius inscribed hypersphere. (b) Diameters through the origin, along each main chord direction. Each diameter is split into a pair of opposite radii shown above and below the axis. The dark magenta sections are symmetrically mirrored, so that any light colored excess shows the deviation from perfect centering of the origin along that direction.

For context, note that for regular polytopes, this ratio remains close to 1 although slowly increasing with the dimension count N. For hypercubes, the value is exactly 1, for dimension counts that are multiples of 4. Otherwise the ratio is in the range of 1.2–1.3 up to $N = 70$. For simplexes, the corresponding mean value range is 1.3–1.5, reflecting the fact that they are slightly more 'pointy' than hypercubes, but still quite isotropic.

By contrast, the mean aspect ratio of 8.1 in the case of the example SSK reflects a far more anisotropic shape. The maximal aspect ratio, the one between the largest and smallest main chords, shows an even more severe contrast by reaching the value 288, in comparison to the respective maxima of about 3 for hypercubes and 6 for simplexes.

This maximal ratio qualifies for being flattened according to the criterion set in Section 7.2.2 and if done, the maximal ratio reduces to 69 and the mean value to 5.3.

The anisotropy of the RBC model SSK is not exceptional. This property is exhibited by all metabolic models investigated so far, and is in fact often more extreme; for example, the Geobacter model for which main chord lengths are shown in Figure 7.7, the mean value is 14.6 and the maximal value about 3600, even after flattening several dimensions.

The second chart in Figure 7.8(b) shows the corresponding diameters through the final coordinate origin, along the same set of directions as the chords. They are uniformly smaller than the chord lengths, by a factor of about two.

The figure has been constructed to illustrate the extent to which centering of the origin has been achieved. To this end, the radius to the SSK bounding hyperplane in each of the two directions aligned to the corresponding chord vector, is respectively shown as the bar length above and below the line. For perfect centering along that direction, these radii would be equal; where they are not, the excess is shown as a light coloured extension above the bar.

It is clear at a glance that the origin is located well away from the boundary in all directions, which is the main goal of centering. There are some visible deviations from perfect centering, as is inevitable for an irregular shape. Although some of this may in principle be due to imperfect centering, tests with regular polytopes do not show such deviations.

Hence, it is plausible to take the deviations displayed, as an indication of how irregular the SSK shape is.

Again, a quantitative measure of the non-centrality is displayed below the figure. Equation (3.9) was used to calculate this, but here extended over the full set of periphery point pairs obtained during centre refining. Typical SSK non-centralities vary between 10% and 30%.

Chapter *8*

Representative FBA Solutions Derived from the SSK

The overall goal of this work is the presentation of metabolic flux values compatible with a constraints-based model, at a level of detail that is intermediate between just the single solution delivered by a straightforward linear programming (LP) calculation, and the comprehensive but intractable description obtained from extreme pathways or elementary mode analysis. So, it aims at finding an economically defined collection of fluxes that, while not complete, is still **representative** of the full range of solutions.

The solution space kernel (SSK) is by construction a subset of all solutions, chosen to directly exhibit the most salient limitations imposed by the optimised objective value and principles such as mass conservation and thermodynamics. For quite a surprising number of models, the SSK turns out to have just single digit dimensions and a similar number of constraints. In such cases, the SSK may itself be considered to be the desired representative set. It has the further benefit that almost all other solutions can be recovered from it by adding multiples of a comparatively small set of ray vectors.

Focusing on the representative aspect, ray vectors define flux combinations for which the metabolic model fails to include realistic restrictions, since it would allow infinite values for them. In that sense, the SSK on its own can be seen as representative of all *realistic* flux values. Note that flux combinations that correspond to ray directions are not necessarily totally excluded from the SSK. They are indeed excluded where coincidence capping was done, but not so for tangent capping. In the latter case, the capping radius puts a limit on ray directions based on other flux combinations that remain finite, as embodied by the feasible bounded facet (FBF) set.

One way to use an SSK as a representative sample would be to calculate all its elementary mode vectors (in principle, all its vertices). This would be viable for a low-dimensional SSK that might yield a few hundred or perhaps thousand EMs. Supplementing these with the set of order N ray vectors produced by the SSK reduction, a set of basis vectors to span the solution space (SS) including points outside of the SSK is obtained.

Note that in using such a basis, only convex combinations of EMs are to be made. But ray vectors are unit vectors so that arbitrary large multiples are needed to reach some points in the unbounded SS. So, ray vector coefficients are still required to be positive but do not need to add up to unity. It was emphasised in Chapter 4, Sections 1 and 5, that the set of ray vectors included in the ray matrix for this work is vectorially complete but not convex complete. This means that recovery of all SS points is not in principle guaranteed by using this basis. However, the careful construction of the ray vectors to be as peripheral as possible ensures a very wide coverage of the SS. The fact that anecdotally it is rare to find additional rays during FBF search, or a deconstruction residue as discussed in section 8.1, support the assertion of a wide coverage.

It is suggested here that in cases where the described basis is compact enough to be viable, it may serve as a manageable, approximate version of the full elementary mode basis for modelling metabolic states.

In the case of models such as the *Geobacter* example where the SSK turns out to have $N = 50$ or more dimensions, it still remains rather cumbersome. Even though its H-representation may be more compact than the full SS by a factor of a hundred, full appreciation of its geometry and shape would still tend to get bogged down by the proliferation of vertices in multiple dimensions that is the bane of elementary mode approaches.

This chapter shows that it is possible to still further reduce the specification of a representative set in such cases. In doing this, the emphasis is on maintaining the *representative* aspect while sacrificing the full recovery of the complete SS by means of ray vectors. Instead of the full SSK, subsets of this are extracted that are meant to cover a *central region* of the SSK.

The issues that arise in this endeavour is to characterise how central the region is and how extensive the coverage of the SSK that is achieved. Section 8.4 addresses these issues, while Section 8.5 shows that the coverage estimation can also be expoited to estimate the overall size of the SSK.

8.1 Validation of the SSK by Deconstruction

In concept, the SSK is merely a partitioning of the SS and there is an exact prescription to recover the full SS from the D-specification of the

SSK as given by Eq. (1.12). The **validation** introduced in this section is aimed at testing if, or to what extent, this full recovery applies to an SSK specified by a given D-specification.

That is relevant because the calculation of the SSK for a particular model may or may not be exact, depending on whether approximations needed to be made for computational efficiency. The most significant of the approximations encountered, is that determination of the base level feasible bounded facet (BFBF) set that is needed to determine tangent capping radii, may have terminated before reaching all members. Also, some feasible fluxes may be excluded if the SSK was flattened to eliminate supposedly insignificant dimensions. In addition to these, several of the algorithms employed random sampling, which means that repeated runs generally produce slightly different SSK specifications. Although the SSK was from the outset understood not to be unique, the question arises whether the final result still measures up to the definition given after Eq. (1.12).

Note that none of the approximations mentioned, compromise the feasibility of points that belong to the SSK; the approximate SSK, even if capped at a radius that ignores some FBFs and/or is subsequently flattened, is guaranteed to be a subset of the full SS and remains representative in that sense. The issue is merely whether all SS points can still be recovered from the approximate SSK by ray addition.

This question is examined by a deconstruction procedure, which is given a flux point P in the SS, and uses just the final D-specification (regardless of how it was derived) to split P into three contributions:

- A vector Q that belongs to the SSK
- A vector R that belongs to the ray space
- A residue vector T

To perform the test, the L1-norm of T is minimised. If a solution exists where this minimum residue is zero, the SSK conforms to the definition at least for the trial case P. Full validation would in principle require this test to be performed for all points in the SS. This is not practically possible, but the extent (if any) to which the calculated SSK falls short may be estimated from the minimised norm values of T, for a selection of trial points P.

The deconstruction procedure can be formulated as an LP minimisation problem. A technical complication is that the constraints that determine Q and R are formulated in different spaces.

Vector Q is defined by the m polytope constraints in n dimensions that form part of the SSK D-specification of Eq. (1.12), and here indicated by the notation C^{mn} for the appropriate constraints matrix. On the other hand, R is a vector in the N dimensions of SS, determined by a different constraints matrix C^{MN} and specified relative to a different coordinate origin. In order to combine them, Q has to be uplifted as in the first equation of (1.11).

In fact, a large gain in efficiency is made by transforming both vectors to the intermediate dimension, reduced solution space (RSS) that was introduced in Chapter 4, Section 2. Since there are no approximations involved in eliminating known fixed values, linealities and prismatic rays by projection, any point that is feasible in the full SS is obtained from a feasible point in RSS by reversing the reduction. Hence it is enough to establish that all feasible points in RSS may be obtained by a zero-residue deconstruction to confirm the SSK validation. This means that the dimension count N now refers to the dimensions of the RSS.

From these considerations, the SSK is taken to be specified relative to the RSS by the data structure

$$D = \{\{C^{mn}, V^m\}, \{O^N, B^{nN}\}\} \tag{8.1}$$

where superscripts are used to indicate the dimensionality of matrices. Using a similar notation I^{NN} to indicate an $N \times N$ identity matrix and 0^{mN} for an $m \times N$ zero matrix, the LP formulation of the deconstruction is then obtained in an extended vector space formed by joining Q, R, T and the auxiliary vector Z needed for L1-norm minimisation. This results in the LP constraint inequalities given by

$$
\begin{pmatrix}
C^{mn} & 0^{mN} & 0^{mN} & 0^{mN} \\
0^{Mn} & C^{MN} & 0^{MN} & 0^{MN} \\
(B^t)^{Nn} & I^{NN} & I^{NN} & 0^{NN} \\
0^{Nn} & 0^{NN} & I^{NN} & -I^{NN} \\
0^{Nn} & 0^{NN} & -I^{NN} & -I^{NN}
\end{pmatrix}
\cdot
\begin{pmatrix}
Q \\
R \\
T \\
Z
\end{pmatrix}
\begin{matrix}
\le \\
\le \\
= \\
\le \\
\le
\end{matrix}
\begin{pmatrix}
V^m \\
0 \\
P - O^N \\
0 \\
0
\end{pmatrix}
\tag{8.2}
$$

In Eq. (8.2), each of the five lines in the matrix symbolically represents a block of constraints, namely:

(1) Ensures that Q is a feasible point in the SSK polytope
(2) Ensures that R is a ray vector in the RSS

(3) Defines the deconstruction expressed by the uplifted Q vector as
$$P = (O + B^t \cdot Q) + R + T$$
(4) Expresses the constraint $Z \geq T$ on the auxiliary vector Z
(5) Expresses the constraint $Z \geq -T$ on Z

This amounts to $(3N + M + m)$ constraints on $(3N + n)$ variables, which demonstrates the benefit of reducing N by performing the deconstruction in the RSS. The appropriate LP objective vector for minimising $\Sigma|T_i|$ is given by $(0, 0,\ldots O_{n + 2N}, 1, 1,\ldots 1_N)$.

The most obvious application of the validation test, is to flux balance analysis (FBA) solutions that were obtained independently of the kernel reduction. Two such solutions are immediately at hand: the FBA fluxes calculated in the original metabolic model, with and without artificial upper flux bounds, and as discussed for the Geobacter example in Chapter 1, Section 3. This could be extended, for example, by recalculation with different LP algorithms.

In such calculations, especially with artificially large flux bounds, the flux vector obtained from standard FBA calculations can have a very large length (sometimes unrealistically so). This needs to be taken into account to judge the significance of any non-zero residue $|T|$ obtained in the validation test. That is best done by comparing the reconstructed vector $(Q + R)$, omitting the residue, with the original vector P. Any percentage mismatch of the vector lengths (i.e., the total flux they represent) and any angle by which $(Q + R)$ deviates from P, serve as indicators of the success of the deconstruction, and which are straightforward to compare between SSK calculations and even for different metabolic models. The result of this test is routinely reported as part of the SSK calculation.

In practice, it is very rare to find any noticeable discrepancy for either of the input FBA fluxes, whether a flux mismatch or a misalignment, for an unflattened SSK. That indicates that failure to guarantee the inclusion of all FBFs, does not anecdotally compromise SSK validity.

However, flattening generally does introduce some discrepancy. An example is the *Geobacter* model, which, as remarked before, yields a very high aspect ratio SSK with $N = 56$ dimensions, but zero discrepancy. Flattening out six thin dimensions reduces the dimensions to 50 and the mean aspect ratio to 14.6, at the price of introducing a 0.5% mismatch in flux value and a 2.5-deg misalignment of the reconstructed flux vector. Such results are useful to judge how much flattening can be done without excessive loss of accuracy.

The validation test is in fact also useful to make a general estimate of the unreachable SS margin introduced by flattening. For this use, the comparison is between a sample point P on the periphery of the unflattened polytope and its deconstruction using a point Q that belongs to the flattened polytope. The residues T are calculated for all the endpoints of the main chords of the unflattened polytope. This choice of sample points is made because it covers a pair of points along each of a complete set of orthogonal directions, and moreover, at least one of each pair is located at an extreme, namely a polytope vertex, so is likely to represent a maximal residue. The mean value of all the calculated residues is taken as a percentage of the mean chord length of the flattened polytope, to give an indicative measure of the relative **flattening error**.

For the abovementioned *Geobacter* model example, the flattening error calculated in this way is 0.1%, in reasonable agreement with the discrepancies found for the FBA flux trials. Again, it is a useful number to inform decisions about how much flattening is acceptable.

Note, however, that the flattening error is not an estimate of the accuracy of flux values in the flattened SSK. All such flux values remain exact feasible fluxes. Instead, it indicates the shortfall in the subset of fluxes that can be constructed by adding rays to flattened SSK fluxes, compared to those at the extremes of the unflattened SSK. In that sense, it indicates the size of the unreachable margin introduced by flattening.

8.2 Discrete Sets of Representative Fluxes

Quite apart from dimensionality considerations, the SSK specification by constraint equations in a multidimensional flux space may still appear too abstract for general use. It would then be expedient to have explicit representative flux values that can be considered representative of the entire range of solutions to the FBA metabolic model.

An extreme case of this is where just a single flux is chosen to represent the 'typical' metabolic state of a cell. It is proposed that a flux point that is *central* in the SSK is a good candidate for this role. Determination of such a point has been a key step in the SSK calculation, and is explicitly given as the origin vector O in the D-specification of the SSK according to Eq. (1.12).

It may not be apparent why the centre of the SSK would be particularly typical. In the more generic concept of the SS as the region of flux space that satisfy physico-chemical constraints, individual cells have metabolic states represented by a 'cloud' of flux points that are distributed in a way determined by environment variables, phenotype and so on. This distribution would generally be expected to extend over only a small fraction of the SS, and may well be located far from the centre. However, as pointed out in the discussion of Chapter 1, Section 1, the term SS is here used in a more restricted sense, perhaps better described as an objective space (OS). The objective, or perhaps multiple objectives, are meant to describe all the overall biological imperatives that determine the metabolic state of a cell. Optimising the objectives would indeed, in general, drive the flux point to an extreme within the generic SS, as is explicitly done when the Simplex algorithm is used for such optimisation. But what remains after optimisation, is the OS, which is the (perhaps small) subset of flux space that contain all the points that are completely equivalent regarding the optimised objective values.

In the OS, therefore, all the aforementioned biological imperatives are already satisfied, and it is plausible to conjecture that the cloud of metabolic points is homogeneously distributed over its extent. This presupposes that it is indeed possible to encapsulate all biological imperatives in a set of linear objectives, but that is a basic tenet of the FBA approach to constraints-based analysis.

Even if it is only approximately true, taking a centre point O as the best overall representative point is analogous to taking the median point as a representative in the standard statistical characterisation of one-dimensional variable distributions. In both cases, it is at least a reasonable starting point in the absence of detailed knowledge of the distribution. By contrast, direct FBA solutions are by construction (at least for Simplex LP) located at a random vertex of the OS, so decidedly atypical. At the very least, the SSK centre point is an improvement on that.

To get a better understanding of metabolic states than is possible from a single typical flux, the next step is to find a set of representative points that can indicate the range of variation about the centre that is allowed by a model. The SSK calculation also presents a ready-made set of such points, namely the periphery points that were used for centre refinement.

The procedure described in Chapter 3, Section 2.4, identifies three distinct subsets of periphery points, all in diametrically opposed pairs, that are combined.

The first of these is the set of endpoints of the main chords, and which also define the directions of the coordinate axes as explicitly listed as the rows of matrix B in the D-specification of the SSK. Each such pair of chord endpoints defines the allowed range of variation of fluxes along an independent (i.e., orthogonal) direction, and were chosen to give the maximal such range. The chord endpoint pairs are not diametrically opposed, so the set is doubled in size by adding the point that is diametrically opposed through the currently chosen centre point O for each endpoint, before incorporating them in the set of periphery points.

The second subset is the constraint vector directions. These are not mutually orthogonal, but they do form a complete basis for the flux space, so on their own also give an account of all the directions along which variation is possible. From the point of view of being representative, they have the benefit of reflecting the shape of the SSK polytope in the sense that each bounding hyperplane is explicitly represented. Again, each intersection point of a constraint vector and boundary hyperplane is supplemented by adding its diametrically opposed partner.

The third subset is formed by a set of directions that are uniformly spaced in hyperangle space, but otherwise randomly chosen. The diametrically opposed intersection points of each direction with boundary hyperplanes defines a pair of periphery points associated with that direction. Once more, this set by construction fully span all SSK dimensions on its own.

Because the three sets are independently constructed, it can sometimes happen that some members nearly coincide and, in such cases, near duplicates are removed.

The combination is a set usually counting between $3N$ and $4N$ points, arranged in pairs. Each pair typically expresses the range of variation along one direction, while remaining approximately central along all other directions by virtue of passing through the SSK centre. A small selection of polytope vertices – those that define the overall shape of the polytope by being located on a main chord – are included, but otherwise, the periphery points are on diameters, spread around the centre to delimit a central region of the SSK polytope.

Examples of the periphery points set for various irregular polytopes in 2D were shown in Figure 3.1. Compared to higher dimensional cases, these may exaggerate how densely periphery points are distributed. Nevertheless, even in high dimensions, the combination of the three point sets as described, aims to ensure that:

- There is at least one periphery point on each boundary hyperplane.
- There are more periphery points on those boundary hyperplanes that span larger hyperangles.
- Overall, the mean number of periphery points on each boundary hyperplane is two to three since the typical number of constraint hyperplanes is around 1.5 N.

It is contended that this set of periphery points is a useful discrete sample of flux values, guaranteed to all be feasible while also exploring the extent of flux variation allowed by the metabolic model. They obviously do not represent the extremes of variation allowed in some particular directions; only if all polytope vertices were included could that be done. By construction, the centre point O is the geometric centroid of all the periphery points. So, for each direction represented by a periphery point pair, the pair is located as closely as possible equidistant from the centre on opposite sides. This is why they are described to define a central region. Such central regions for some 2D polytopes were also displayed in Figure 3.1.

As an example of practical use, consider a case where it is necessary to evaluate a mean value of some function $F[x]$ of metabolic flux over all fluxes x allowed by the model. Then, a discrete mean over just the periphery points (and possibly the centre point O) could be used as an easily calculated estimate.

If F is a linear function, it would be equivalent to simply evaluate $F[O]$, because O is by construction the centroid. This serves as a mathematical justification for sometimes using O as the sole representative flux.

But for non-linear functions, the effects of the allowed variation would be reflected in the mean value obtained from including all periphery points.

Another use of the periphery points is as a final check whether the flux variables that have been identified as remaining fixed over the extent of the SS, is complete. For this purpose, the periphery points are transferred back as vectors in the full flux space, and any component that has the same value for all periphery points is tentatively identified as fixed. This classification is confirmed only if the corresponding component also has the value zero in all the ray vectors, to ensure that it cannot change when fluxes outside the SSK are constructed by adding rays. Most of the fixed values determined in this way coincide with those already found before, but in some cases, several additional fixed fluxes can be added to the known list.

8.3 A Representative Flux Region: The Peripheral Point Polytope

The convexity of the SSK, guaranteed by its half-space specification, means that any convex combination of peripheral point fluxes is also a feasible flux. Hence the set of periphery points in fact defines a continuous region of feasible flux values as is formally given by the convex hull of the periphery points.

This feasible region can be conceived as a polytope with the periphery points as its vertices. Actually, not all periphery points need to be included; those that are themselves representable as convex combinations of others can be omitted without loss. The remaining periphery points are in fact the vertices of the convex hull.

In the 2D case shown in Figure 8.1, finding the convex hull would allow the feasible region to be represented by just seven periphery points out of the original 10. Such reduction is, however, rather unlikely in

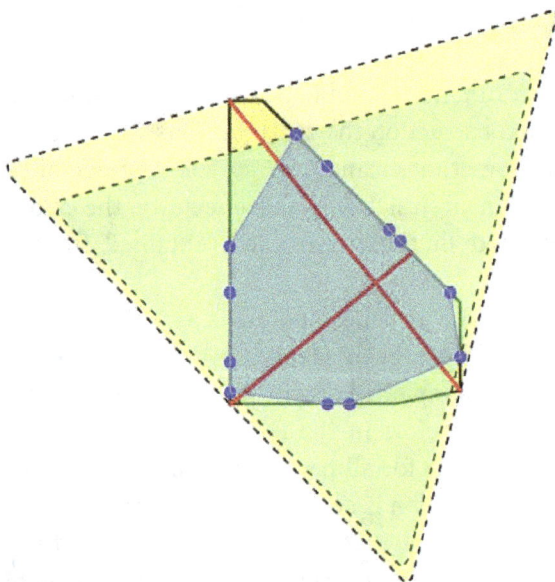

Figure 8.1   An irregular 2D polygon (solid lines) and its inscribed PPP, based on the calculated periphery points (blue). The main chords are shown in red. The yellow and green dashed triangles respectively circumscribe the full polygon and the PPP.

higher dimensions because with only a small number of periphery points on each boundary hyperplane (itself still having multiple dimensions), it is quite unlikely that sufficient subsets of them will fall on the same line, plane and so on in order to allow omission. So, it suffices to take the central feasible region as the polytope V-specified (in the terminology of Chapter 1, Section 2) by all periphery points.

This region is designated as the **peripheral point polytope** (PPP), and can be considered to be a *central region* of the SSK by the same arguments applied to the set of periphery points above; in particular, it contains the origin point O as the centroid of *all* its vertices.

It is interesting to compare the full SSK and the PPP. Both of these are polytopes in the same number of dimensions. The first is given by an H-specification, and the second as a V-specification. Depending on the SSK dimensions, the number of SSK vertices may be very large, making its V-specification unmanageable; but the vertex count of the PPP is by construction only a low multiple of the number of dimensions. This makes it quite manageable as a V-specification, but raises the question whether the PPP can realistically have sufficient coverage of the SSK to be a representative region.

The most obvious way to consider coverage is to compare the hypervolumes of the PPP and SSK, that is, to define the **coverage** as the hypervolume fraction achieved by the PPP.

Figure 8.1 shows an example of an irregular 2D polytope and one possible PPP, together with some enclosing triangles that will be discussed further below. For illustrative purposes, the first subset of periphery points normally obtained from the main chords were omitted to obtain a visibly incomplete coverage. Even so, it seems visually evident that a reasonable coverage is still attained.

But the examples discussed in connection with Eq. (1.6) make clear that it would not be realistic to expect that to extend to high dimensions. For example, even though in 2D, the inscribed circle covers 79% of the area of a square, the hypervolume ratio of the inscribed hypersphere to that of the hypercube diminishes exponentially with dimension count N and becomes negligible for a value as small as $N = 10$. The PPP is an inscribed polytope, and might be expected to fare even worse than an inscribed hypersphere.

To avoid this incidental consequence of how hypervolumes scale with dimensions, it is proposed that a ratio of representative *diameters* rather than hypervolumes is used as a measure of coverage. Exactly how the diameters are defined and calculated is to be clarified below.

Assuming that the PPP does provide acceptable coverage of the underlying SSK, it is noted that the approximate basis representation of the SSK that was proposed in the introduction of this chapter could be even further simplified. Since the PPP is already in the *V*-representation, the calculation of elementary modes for the SSK can be skipped and the flux vectors at the vertices of the PPP directly used instead. This obviously introduces further approximation, but may be adequate to give an overview of fluxes that are feasible according to the metabolic model, while remaining manageable even for a comparatively high-dimensional SSK.

8.4 Estimation of the PPP Coverage

This section discusses two approaches to estimate the extent to which the PPP covers the entire SSK polytope.

In the first method, sampling yields a probability that a flux point randomly sampled from the SSK falls in the PPP. This probability in essence reflects the hypervolume ratio of the two polytopes, although it can be reduced to a diameter ratio. As such the hypervolume scaling discussed previously means that sampling estimates can only be successfully applied in cases with a small dimension count. For such cases, it is useful to give a benchmark for the second method.

The second method compares the diameters of two geometrically similar envelopes that respectively circumscribe the PPP and SSK. With appropriate choice of the envelope shape, this ratio can be efficiently calculated in far larger dimensions. It is a useful general purpose coverage measure, and interpretation of its value is facilitated by comparison to the values obtained by sampling in low-dimensional cases where both methods can be applied.

8.4.1 *Coverage Estimation by Sampling*

Although the concept of approximating an inscribed hypervolume fraction by the observed sampling frequency is self-evident, taking random samples from higher dimensional polytopes is not so straightforward, especially for the high aspect ratios typical of SSKs.

There are three aspects to address: how to sample from the SSK, how to decide if a given sample point belongs to the PPP and when to stop polytope sampling.

A simplistic way to sample the SSK would be to take a random flux point with components within the bounds given by the SSK bounding box (as described in the FVA [flux variability analysis] discussion of Chapter 1, Section 3), and rejecting the point if it does not satisfy the known feasibility constraints that define the SSK as its H-representation. However, this is prohibitively inefficient since for all but the lowest dimension counts, the vast majority of sample points will be rejected because as argued before, the SSK occupies a negligible fraction of the bounding box.

Uniformly sampling high-dimensional polytopes is still an active area of research, and elaborate mathematical procedures have been proposed for it, including in the field of metabolic flux calculations [1–2]. For the present limited purpose, the following pragmatic approach is however deemed sufficient.

The method makes use of the fact that given a direction in N-dim space, the offcut radius to the periphery in both forward and backwards directions is quick to calculate, as outlined in Chapter 3, Section 2.4. The basic strategy is to choose a series of random directions, and to find a predetermined number of random points along each direction. These points are kept inside the polytope by taking each as a randomly chosen fraction ϕ of the known radius R to the periphery, that is, located at a radius $r = \phi R$ along the sample direction.

This radial sampling mode means that each sampled point in N-space represents a volume element in hyperspherical coordinates, and which is proportional to r^N in N dimensions. Hence, to produce a uniform spatial sampling density, the number of sample points needs to accordingly increase with radius by a power law. This is achieved in the standard way by using the inverse of the desired distribution function, that is, a random fraction Θ is sampled uniformly from the interval $(0, 1)$ and then transformed as $\phi = \Theta^{(1/N)}$.

Next, the number of points to be sampled along each direction is to be determined. As the periphery radius R along each direction may be different, this is done by choosing a standard radial span S to be associated with a single sample point. Then, the number of sample points for a given direction is chosen as $n = R/S$. That gives a linear increase of the sample count with R, instead of the power law increase strictly required. This is done because there is a price to pay for having multiple sample points along a single direction, namely the loss of randomness that result from their being collinear.

The extent to which randomness is sacrificed depends on the choice of S. At one extreme, S could be chosen as the smallest peripheral radius for the polytope. In this case, most other directions will have multiple sample points along each direction, the more so with high aspect ratios, and the loss of randomness can be large. At the other extreme, if S is chosen as the largest peripheral radius, the value of n would be less than one for other directions. Such a fractional value is interpreted as a probability: for example, if $n = 0.5$, a sample point will be taken along this direction for only 50% of such sampled directions. Because there will always be exactly one sample point along any direction that survives the rejections implied by the probability choice, there is no loss of randomness for this choice of S. However, it suffers from inefficiency because many sampled directions are rejected without contributing any sample points.

As a compromise between efficiency and randomness, the final choice is to take

$$n = R/S; \quad S = Median[R_i] \tag{8.3}$$

where index i enumerates all the forward and backwards directions associated with the polytope main chords, that is, the SSK coordinate axes.

Eq. (8.3) ensures that in directions where the polytope is elongated more sample points are chosen than along thinner directions, but only moderately so. The shortfall from a power law increase for a single direction, is however compensated by appropriately making the next choice: the random sample directions.

A purely random choice of direction could be made by choosing each component of the N-vector as a random number in a fixed interval such as $\{-1, 1\}$. Then by choosing a larger number of directions along the longer axes of the polytope, the density of sample points along these directions is further increased beyond the linear growth built into Eq. (8.3). It is easily done at virtually no computational cost, by simply choosing the sampling interval for each component of the vector as the interval $\{-R_i, R_i\}$. When the resulting vector is renormalised to a unit vector, its components along the longest axes are on average larger than along shorter axes. This means that in repeated direction sampling, the vectors tend to be bunched along long axes of the polytope and thinned out along the short axes.

This strategy is in fact equivalent to scaling each orthogonal direction by the reciprocal of its radius, so that the SSK transforms to one with equal aspect ratios. That can be seen numerically by multiplying the set of directions by the diagonal matrix of scale factors and observing that they then achieve a uniform distribution. The effect achieved is very similar to the 'rounding' strategy advocated by De Martino et al. [2] for sampling from high aspect ratio polytopes.

It is found that the sampling strategy described in this section achieves a sufficiently uniform spatial distribution, for example, for highly asymmetric polygons and polyhedra that the next issue can be addressed: given a sample point in the SSK polytope, does it fall inside the PPP?

This can be answered by considering that the PPP is a convex polytope, of which the vertices are known. Any point inside the PPP can be expanded in the convex basis formed by the vertex vectors, with coefficients that are between zero and one, and which add up to one. Points outside the PPP may need some coefficients to be larger than one, and even if not the sum of coefficients is larger than one.

Hence membership of the PPP can be tested by using LP to find the nonnegative coefficient, vertex point expansion of a sample point, and which minimises the coefficient sum. If this minimum is larger than one, the point is outside the PPP.

So, the coverage calculation amounts to generating sample points as described; for each point an LP establishes whether it is inside the PPP, and the fraction of the sampled points found inside is the current estimate of the coverage. Sampling proceeds by adding sampling directions, until the estimate is considered sufficiently converged to be terminated.

To make this decision, consider the coverage calculation as a series of Bernoulli trials that uses the observed frequency of successfully finding a point that belongs to the PPP, to estimate a confidence interval for the actual probability (and hence hypervolume coverage of the PPP). This is a standard statistical problem of finding a binomial proportion confidence interval, for which there are several formulas in common use. We choose the 95% Jeffreys confidence interval that has the advantage that it is symmetric in the sense that the probabilities that the interval lies above or below the true value are both close to 2.5%. According to this formula, given x successes in n trials, the confidence limits are the 2.5% and 97.5% quantiles of a Beta distribution with parameters $(x + \frac{1}{2}, n - x + \frac{1}{2})$.

Based on this, the sampling of points is continued until the width of the confidence interval has decreased to a preset maximum, taken as the value 0.05. The corresponding limits give the 95% confidence coverage estimate.

As a simple test of the procedure, a 2D square and a three-dimensional (3D) cube (with, in each case, the centre points of all sides taken as the periphery points), yield respective PPP coverages of 50% and 16.5% by simple geometry. For the square, a typical random sampling run terminated after sampling 664 directions and 1534 points, to give the 95% C.I interval as (46, 51) %. In the 3D case, 296 directions and 743 points were sampled to give the interval estimate as (12, 17) %. Flattening the square or cube to an aspect ratio of 25, very similar results are obtained. Less symmetric examples in 2D and 3D also give similar results. Hence, the sampling method is considered a viable method of coverage estimation, at least in low dimensions.

8.4.2 *Coverage Estimation by Similar Envelopes*

The number of sample points needed to establish the coverage with reasonable accuracy can be quite large (even in two or three dimensions), and each point requires an LP calculation to classify it. To alleviate this problem, an alternative approach to estimate the coverage in a geometric way is investigated in this section.

The main idea is to compare the sizes of geometrically similar shapes or *envelopes* that circumscribe the SSK and PPP respectively. One might, for example, test by what factor the PPP has to be enlarged in order to enclose the SSK, or the reverse, the factor by which the SSK would need to shrink to enclose the PPP. Either of these would be rather biased, in comparing the size of a shape that perfectly covers one polytope with one that only partly covers the other.

To remove such bias, one could do the comparison using a shape that is unrelated to either, such as the respective circumscribed hyperspheres centred at the origin. There would be an implicit 'coverage' assumption, that each of the polytopes occupies a similar fraction of their respective envelopes, so that the ratio between the envelopes is a good approximation of the ratio between the polytopes.

A hypersphere would not generally fulfil this assumption, for example, if the two polytopes have different aspect ratios. Also, it would in general only pass through a single point of the polytope periphery (a vertex).

It would be better to choose an envelope shape that can be broadly adapted to the aspect ratios of the polytopes, and more closely follow the periphery by sharing multiple points with the polytope it contains. Also, we should allow its centre to shift so as to keep the envelope as small as possible.

Even so, there might be some coincidental bias if the chosen shape fits one of the two polytopes better. This cannot be eliminated, but it can be counteracted by repeating the calculation for multiple spatial orientations of the chosen envelope shape and taking an average value over all of them.

The choice of a regular simplex as the envelope shape provides these advantages in addition to being straightforward to calculate. It is in some ways like a discretised hypersphere, in the sense that there is a unique centre (the centroid of its $(N + 1)$ vertexes) from which the distances to all vertices are the same, and also the perpendicular distances to all boundary hyperplanes are the same. In other words, this centroid is the centre of both the circumscribed and inscribed hyperspheres of the simplex, and which respectively pass through *all* vertexes or touch *all* sides of the simplex. Moreover, the hyperangles spanned at the centre by all vertex pairs are the same, that is, the vertices of a regular simplex are uniformly distributed over the hyperangle space.

The 2D example shown in Figure 8.1 illustrates the essential idea. A major and minor equilateral triangle that encloses the SSK and PPP respectively, are shown in yellow and green. Their sizes can be compared as a ratio of areas, or alternatively as a ratio of any suitable distance measure since the triangles are similar.

The calculation in N dimensions proceeds by finding a major regular simplex that encloses the SSK, and a minor one with the same orientation that encloses the PPP. Comparing the radii (either to vertexes or to sides) of the major and minor simplexes gives a measure of their relative scales, and is repeated for multiple orientations.

With this in mind, the following steps are required:

(a) Choose the i-th uniformly spaced set of $(N + 1)$ directions, as the vertex directions of a randomly oriented regular simplex.
(b) For each direction j of this set, find the distance R_{ij} from the coordinate origin to the closest hyperplane that defines a half-space containing the SSK polytope (and hence also the origin, which is by construction inside the polytope).

(c) Calculate the common perpendicular major radius R_i to the resulting $(N + 1)$ boundary hyperplanes that form the envelope regular simplex, as measured from the simplex centroid rather than the origin. The calculation method is described below.

(d) Repeat steps (b) and (c) for the PPP polytope to find the minor radius r_i.

(e) Repeat steps (a) to (d) for a chosen number of random orientations i, and take the following ratio ρ as a measure of the relative scales of the envelope sizes:

$$\rho = \frac{\text{Mean}[r_i]}{\text{Mean}[R_i]} \qquad (8.4)$$

Note that by construction, each simplex boundary hyperplane will pass through at least one vertex of the relevant polytope, so the simplex shares at least $(N + 1)$ points with the polytope, and will broadly reflect the different aspect ratios of the polytope.

Steps (b), (c) and (d), performed for a fixed orientation i, need further clarification.

Since the SSK polytope is only known as its H-representation, step (b) requires an LP optimisation for each direction. Given the direction as the unit vector $\hat{r}_{ij}$, the orthogonal hyperplane located at a distance R_{ij} from the origin, is given by all points x that satisfy $\hat{r}_{ij} \cdot x = R_{ij}$. Shifting the hyperplane outwards as far as possible, such that there exists an x that falls inside the SSK polytope, gives the desired location as described by (b). So, R_{ij} is determined by an LP maximisation with the objective vector $\hat{r}_{ij}$.

For step (d) on the other hand, the PPP polytope vertices are known. It is merely necessary to select the last vertex that intersects the $\hat{r}_{ij}$ hyperplane as it is shifted outwards. For a vertex located at position v, it will lie on the hyperplane defined by $\hat{r}_{ij} \cdot x = R_{ij}$ if its projection along $\hat{r}_{ij}$ equals the hyperplane radius, that is, $\hat{r}_{ij} \cdot v = R_{ij}$. So, calculating the projections of all vertices (i.e., periphery points) and taking the largest value determines the appropriate R_{ij}. If we define a matrix Ψ where each row is a periphery point position, and D as the matrix where row j contains the direction vector $\hat{r}_{ij}$, the projections of all periphery points along all directions are obtained in the single matrix multiplication $\Psi \cdot D^t$. The required set of R_{ij} values is obtained as the row-wise maxima of this product.

The outcome of either step (b) or (d) is a set of $(N + 1)$ distinct R_{ij}-values that define orthogonal distances from the coordinate origin to the respective boundary hyperplanes of an envelope simplex. This is a regular simplex, because these hyperplanes by construction have orientations that are uniformly distributed hyperangles, and so make equal angles with each other. The reason that the R_{ij}-values are nevertheless unequal is that the simplex is not centred at the origin. In fact, the minor and major simplexes may have different centres, as in the example shown in Figure 8.1.

However, step (c) requires the calculation of a single common radius value for each, as measured from its proper simplex centroid. This is easily done. The principle is demonstrated by the 2D case of an equilateral triangle, as shown in Figure 8.2. In the figure, the centroid X divides the triangle into three congruent triangles with equal surface areas AB · $h/2$. The common height h is also the radius of the inscribed circle. Alternatively, an arbitrary point Y divides it into unequal triangles, but as the total surface area is the same, and also AB = BC = AC, it follows that

$$3(AB \cdot h/2) = BC \cdot h_1/2 + AC \cdot h_2/2 + AB \cdot h_3/2$$
$$\Rightarrow h = \tfrac{1}{3}\sum_i h_i \qquad\qquad (8.5)$$

In a similar way, a regular simplex in N dimensions has N congruent $(N-1)$-simplexes as its sides, and its volume can equivalently be divided up either into equal N-simplexes sharing a common height equal to the

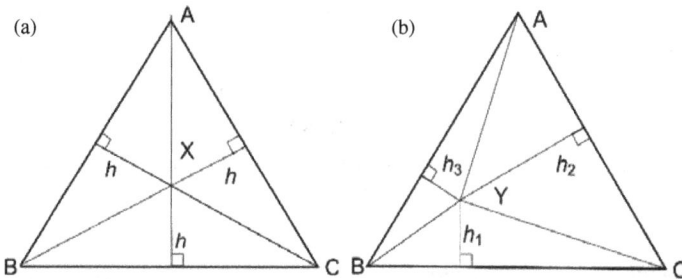

Figure 8.2 Equilateral triangle ABC (a) with centre X dividing it into congruent triangles BCX, ACX and ABX with common heights h and (b) Shifting the origin to an arbitrary interior point Y gives unequal triangles BCY, ACY and ABY with respective heights h_1, h_2 and h_3.

inscribed hypersphere radius, or into unequal simplexes sharing a common apex at an arbitrary interior point Y.

In the case of the calculation described by the previous steps, Y is in effect chosen as the coordinate origin relative to which the SSK is specified. Then, the R_{ij}, for a fixed i, are the perpendicular heights above the simplex sides enumerated by index j, and as in Eq. (8.5), it follows that their mean value is the radius of the inscribed hypersphere for the major simplex in orientation i:

$$R_i = \frac{1}{N+1} \sum_{j=1}^{N+1} R_{ij} \tag{8.6}$$

In step (e), the mean value over all orientations i is taken for each of the major and minor envelope simplexes to give a coverage ratio.

It only remains to specify the range for index i to choose in step (e). The number of possible orientations obviously increases with the dimension count. Nevertheless, trials show that the mean value converges quickly even in high dimensions, so the pragmatic choice is made to limit the calculation to just 10 randomly chosen orientations in all cases.

8.5 Comparing SSK Sizes of Different Models

Although coverage estimates are mainly intended to justify the use of the PPP as being representative of the SSK, knowing the relative size of the PPP would also enable estimating the SSK hypervolume if an absolute value for the PPP hypervolume can be obtained.

The relevance of the SSK hypervolume to interpreting metabolic models is that it could be used to compare the extent to which the constraints and objectives in different models restrict the metabolic state to a limited range. For example, the presence or absence of a particular nutrient may produce a marked difference in the SSK hypervolume. This could plausibly reflect the adaptability of the organism to survive in different environments. Another example might be to evaluate the scope offered by an organism for bioengineering it to produce a desired metabolite. If the SSK hypervolume is small, it implies that the metabolic fluxes are highly determined and do not leave much room for manipulation.

Note that such comparisons can usually not be made for the full SS hypervolume, which is infinite for the vast majority of metabolic models.

It is only by the careful exclusion of ray directions that the SSK becomes finite. Moreover, it is by design bordered by the FBFs that embody actual limitations imposed on flux values by the interplay of physical and chemical rules as mediated by the metabolic network, and regulatory effects represented by the FBA objectives. So, an SSK volume comparison is also more likely to be biologically relevant than a comparison of the complete SSs.

8.5.1 *Measures of SSK Hypervolume*

A step towards volume assessment is to show that a simple formula gives the hypervolume of a major part of the PPP, namely the polytope obtained by taking just the maximal orthogonal chord endpoints (the first of the three subsets of periphery points) as its vertices. This is a special case of all inscribed polytopes with $2N$ vertices, that are paired to define N orthogonal chords. Call these chord-orthogonal polytopes (CHOPs).

The hypervolume of a CHOP can be derived from a well-known formula for the hypervolume of an N-dimensional simplex as a determinant, given by

$$V_{simplex} = \left| \frac{1}{N!} \det \begin{pmatrix} v_0 & v_1 & \cdots & v_N \\ 1 & 1 & \cdots & 1 \end{pmatrix} \right| \tag{8.7}$$

where v_0 , v_1, ...v_N are column vectors defining the positions of the $(N + 1)$ vertices.

Also, note that a simplex S_{N+1} in $(N + 1)$ dimensions is formed from S_N in N dimensions, by adding a single point along a direction orthogonal to the hyperplane that contains S_N. Using this, a CHOP can be built up starting from the 1D simplex formed by one of its chords of length c_1, and then progressively adding dimensions one at a time by adding subsequent chords.

We start from a 1D simplex (a straight line) which has a length (i.e. hypervolume) $V_1 = c_1$. The first step creates two triangles (i.e. 2D simplexes) that each shares the vertices of the 1D simplex as its base. Combining the triangles, their heights add to form the new chord so the area of the 2D CHOP is $V_2 = \frac{1}{2}V_1{}^*c_2 = \frac{1}{2}c_1{}^*c_2$ where c_i are the respective chord lengths. In 3D, there are a total of four tetrahedrons, in pairs that share either their base triangles or their heights leading to a total volume of $\frac{1}{3}V_2{}^*c_3 = c_1{}^*c_2{}^*c_3/6$. Extending this argument to higher dimensions, it is easily seen that that the general formula for the hypervolume of a CHOP in N dimensions is

$$V_{CHOP} = \frac{1}{N!} \prod_{i=1}^{N} c_i \tag{8.8}$$

In principle, the maximal volume CHOP would be determined by maximising the product of chords c_i or equivalently the sum of $\text{Log}[c_i]$. These are non-linear objectives putting the problem in the realm of Geometric Programming (GP) rather than LP, further complicated by the fact that the c_i themselves are already non-linear functions of the flux variables. Such problems are computationally far more challenging than LP's so this does not seem worth pursuing to find a value that is in itself only a partial estimate of the PPP hypervolume.

Instead, the LP calculation of maximal chords described in Chapter 7, Section 3, systematically maximised the c_i one-by-one, while ensuring that all endpoints remain feasible. This can be seen as a kind of greedy heuristic for the maximisation of the chord product given by Eq. (8.8). In many cases, it may deliver the actual maximum hypervolume CHOP that can be formed with $2N$ vertices. But even if not, it should still give a reasonable approximation to it, and at the very least, a lower limit to this volume.

The centroid of the vertices of this CHOP may not be optimally centred for the whole SSK because it only selects a small subset of its vertices. The centre refinement performed subsequently adds additional vertices to achieve better centering of the centroid, and in the process each additional vertex adds volume to the underlying CHOP.

For both reasons, Eq. (8.8) using the LP calculated chord lengths, constitutes a lower limit estimate of the hypervolume of the PPP. Since the PPP itself defines a subset of the SSK, this equation also gives a lower limit for the SSK hypervolume, but may well underestimate it by a considerable factor.

It appears that although an order of magnitude estimate of the SSK hypervolume is possible using the calculated main chord lengths, it is unlikely to give a useful basis of comparison between metabolic models.

This is supported by two further observations. First, as described in the previous section, the volume discrepancy between the SSK polytope and its inscribed objects, whether a hypersphere, the PPP or the CHOP, is very strongly dependent on the dimension count, which generally differs between models. This is a fundamental geometrical scaling effect that cannot be avoided.

A second factor is that, as a consequence of the chord product in Eq. (8.8), the CHOP hypervolume is also sensitive to the presence of even just one negligible chord length, which as pointed out before quite commonly occurs in metabolic models. This can give a negligible hypervolume that is not representative of the substantial range of flux variation along most directions.

As an illustration, in 3D, one can picture an SSK with the shape of a flat polygonal disk. Although the volume of the disk may be negligible, its mean diameter in the plane of the disk would be a better indicator of the flux variation it allows.

The conclusion reached here is that a linear distance measure of the SSK spatial extension is a better way to compare models than hypervolumes.

8.5.2 *Distance Measures of the SSK Size*

Using a *distance* measure instead of a *volume* measure to characterise the size of the SSK avoids the described difficulties caused by dimensional scaling and excessive sensitivity to the SSK shape.

Given a well-centred SSK, its most obvious single distance measure of size is its mean diameter d. This is not easy to calculate directly from its H-specification for large dimensions, but the PPP by construction supplies a sample of diametrically opposed periphery point pairs along representatively chosen directions. Define the arithmetic mean of these PPP diameters as d_{ppp}. The question arises whether d_{ppp} tends to over- or underestimate d.

Overestimation can be expected if most peripheral points are located far away from the centre, such as at vertices, at the expense of nearby points, and the worst bias would be expected for high aspect ratio polytopes. A simple 2D example would be an elongated rectangle in which the four corners are chosen as periphery points. The opposite bias is where the four points are chosen near the centres of the long sides of the rectangle and would produce an underestimation of d.

The procedure for choosing peripheral points during centering was constructed to avoid such bias: the main chord endpoints are at far vertices; an equal number of constraint vector intersections are at the closest points on each side, and a further subset randomly samples equally spaced directions. Nearby points are more likely to be fully represented because all sides are sampled, whereas the fraction of vertices sampled is small and reduces with dimensions, even though the ones that are furthest away are explicitly included by the use of main chords. So, it seems plausible to assume that with the actual choice of PPP vertices, $d_{ppp} < d$.

This assumption is supported by the fact that the coverage ratio ρ calculated from the major and minor envelopes described in Section 8.4.2 is always <1. This means that the SSK contains points that lie outside of the PPP and need a larger envelope to capture them; the corresponding diameters are large and will tend to increase the mean diameter value.

This argument suggests that the extent to which the major envelope radius exceeds the minor one is a measure of the d_{ppp} underestimation, that is, an improved diameter estimate is

$$d \approx d_e = \frac{d_{ppp}}{\rho} \tag{8.9}$$

This estimate is still dependent on the success of the periphery point selection. To limit uncertainty, it is useful to also establish lower and upper limits within which d can plausibly range.

One such putative upper limit is the mean orthogonal main chord length:

$$c_m = \frac{1}{N} \sum_{i=1,N} c_i \tag{8.10}$$

Each *chord length* is by construction the maximal distance along a coordinate axis between two SSK points, and the *diameter* along this direction is smaller because it is additionally constrained to pass through the origin. By the same token, when means are taken over all directions, the diameter along any direction can never exceed the similarly aligned chord, so $d < c$. However, the limited number of directions used in Eq. (8.10) means that c_m is only an approximation to c and so whether $d < c_m$ is less certain.

In fact, the discrepancy between the chord and diameter along the same direction is usually quite large for the highly anisotropic polytopes typically obtained in SSK calculations. Examples discussed in Chapter 7 and illustrated for example in Figure 7.7 and Figure 7.8 show that the diameter can be an order of magnitude shorter than the chord.

This large discrepancy can usually be expected to make up for the lack of diversity of directions, and so Eq. (8.10) is generally a realistic upper limit. The exception is when there are thin directions, in which case c_m can occasionally produce a value even smaller than the d_e estimate of Eq. (8.9), which is based on a much larger set of directions given by periphery point pairs. This anomaly is removed when the SSK is appropriately flattened. This means that c_m can only be taken as a soft upper limit to the mean diameter d.

The regular simplex envelope of the SSK, used to estimate coverage, can also be used to put limits on diameter estimates. The coverage calculation yields the inscribed hypersphere radius R_i of the major simplex. Simple formulas for various other size measures for the envelope can be established by straightforward trigonometry, based on the fact that the angle θ subtended at its centre by each edge of a regular simplex in N dimensions is given by $\mathrm{Cos}(\theta) = -1/N$.

Consider the plane containing two vertices U and V of a regular simplex, as well as its centre point C. This is illustrated in Figure 8.3.

Using the expression for θ, the quantities of interest can all be expressed as multiples of the inscribed hypersphere radius, as follows:

- The inscribed diameter, $D_i = 2\,R_i$
- The circumscribed diameter, $D_c = 2\,R_c = 2\,N\,R_i$

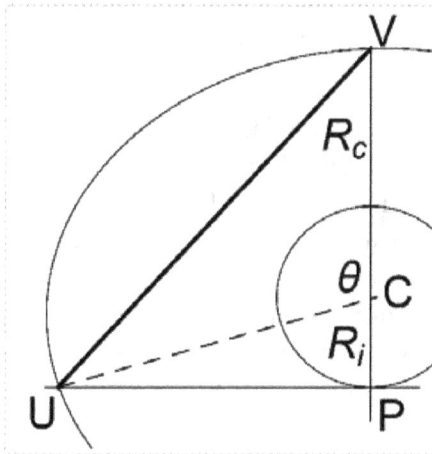

Figure 8.3 A plane intersecting a regular simplex in N dimensions, and containing two vertices U and V and the simplex centre C. The extension of the line VC intersects the simplex side opposite to V, perpendicularly at its centre point P. The distance UV is the maximal chord length of the simplex, CP is the inscribed radius R_i and $CV = CU$ is the circumscribed radius R_c.

- The maximal simplex diameter $D_{max} = R_i + R_c = (N + 1) R_i$
- The maximal simplex chord length, $C_{max} = \sqrt{2N(N+1)} \, R_i$

Here, capital letters were used to symbolise attributes of the enclosing simplex, analogous to the lowercase letters used for the SSK. The values have the fixed sequence order $D_i \leq D_{max} \leq C_{max} \leq D_c$, with equalities only for the trivial case $N = 1$. To establish the location of d in this ordering, note that $d \leq D_{max}$. If not, there would have to be at least one diameter which has endpoints further apart than D_{max}, and this contradicts the fact that by construction all points in the SSK are inside the enclosing simplex.

Regarding a lower limit, the maximal SSK diameter d_{max} must satisfy $d_{max} \geq D_i$ because if not, a simplex envelope with a smaller inscribed diameter would also enclose the entire SSK. That would contradict the fact that by construction, the chosen envelope is the smallest enclosing regular simplex.

To reduce this to a limit on the mean value d, we write $d = f d_{max}$ where f is a fractional value. This value can range between $f = 1$ which is the case when the SSK is a hypersphere where all diameters are equal, to the value $f = 1/N$, which corresponds to the extreme case where all diameters except d_{max} are negligible, that is, the SSK is essentially a straight line. Substituting the latter f value gives the lower limit:

$$d \geq f D_i \geq \frac{1}{N} D_i \tag{8.11}$$

This is a somewhat pessimistic lower limit, as it is based on a hypothetical SSK shape with an infinite aspect ratio. It is possible to refine this somewhat by taking the actual finite aspect ratios that come from the SSK main chord calculation into account, but in practice these still give an f value proportional to $1/N$ with a coefficient not much larger than one. The previously stated heuristic limit $d > d_{ppp}$ is likely a more realistic indication of the SSK diameter range.

To summarise, the softer limits based on the PPP can be combined with the hard limits supplied by the simplex envelope to define a plausible range for the mean SSK diameter by the inequalities:

$$Max[d_{ppp}, \frac{D_i}{N}] \leq d \leq Min[c_m, D_{max}] \tag{8.12}$$

Within this range, the best point estimate of the SSK mean diameter is given by Eq. (8.9). As pointed out earlier, the mean chord estimate c_m tends to underestimate the diameter when some chords are negligible. This can give rise to the anomaly that the point estimate exceeds the upper limit given by Eq. (8.12). In this case, c_m is discarded and the upper limit is taken as D_{max}.

However, the observation of such a conflict is a strong indication that the SSK has thin dimensions that need to be flattened out, and when done, the anomaly is eliminated. This obviously only applies when the chord lengths were calculated by LP. For large models where diameters have been used to estimate chord lengths, the conflict noted can arise more easily and cannot be reliably used as an indication for flattening.

References

1. B. Meister & P. Clauss, Uniform random sampling in polyhedra. (*IMPACT 2020 — 10th International Workshop on Polyhedral Compilation* Bologna, Italy, 2020).
2. D. De Martino, M. Mori, & V. Parisi, Uniform sampling of steady states in metabolic networks: heterogeneous scales and rounding. *PLoS ONE*, **10** (2015) e0122670. https://doi.org/10.1371/journal.pone.0122670.

Chapter *9*

Case Studies of SSK Calculations for Realistic FBA Models

In this final chapter, the considerable variation in the outcome as well as the course that a solution space kernel (SSK) calculation can take is explored by applying it to a number of Flux Balance Analysis (FBA) models in the public domain. These models were mostly taken from the BiGG online depository [1], or directly from the published literature in a few cases.

9.1 Solution Space Classifications

For this discussion, it is useful to introduce a classification of solution spaces. This applies firstly to the SS as specified by the stoichiometric constraints as in Eq. (1.1), as well as any constraints on the value range of flux components, such as the common restriction to non-negative flux values $f_i \geq 0$ that determines a reaction direction. Any upper limits to a flux value dictated by physical laws are included, but *artificial* flux upper limits are specifically excluded. Any constraints assigning a definite value to one or more FBA objective functions are also considered to be part of the SS specification.

As the kernel calculation progresses, the remaining solution space changes and so does its classification. Of particular relevance is the classification of the reduced solution space (RSS) obtained by separating fixed fluxes, applying the stoichiometry constraints and removing prismatic rays and linealities. This determines the course of the more computationally challenging steps towards coincidence and tangent capping before a final compact SSK is achieved.

9.1.1 *Compact Solution Spaces*

In principle, it is possible that all flux components are limited to finite ranges, and the model would then be classified as having a **compact** SS. The vast majority of metabolic models in their original form are in fact open in some directions in flux space once artificial limits are removed, but the imposition of a fixed objective value can sometimes induce a partially open SS to become compact.

A trivial example is a three-dimensional (3D) model with fluxes f_1, f_2 and f_3, all ranging from 0 to ∞. Without an objective, the SS would be an open cone formed by the three positive coordinate axes in flux space. Assigning a fixed value $\phi > 0$ to the flux combination $f_1 + f_2 + f_3$ reduces the SS to the triangle formed by the intersections of the plane $f_1 + f_2 + f_3 = \phi$, with the coordinate axes. This triangle is a compact SS since it has no open directions.

Another way that a compact SS can arise is when open directions (i.e., ray directions) are eliminated by projecting out prismatic rays or by coincidence capping, as used in reduction to the progenitor hyperplane. As the end goal is to construct a compact kernel, the SSK calculation is terminated whenever a compact SS is reached in an intermediate stage.

To confirm that the SS is compact, it needs to be shown that it has no ray directions. Ray finding is the first step in the SSK calculation, whether by standard matrix operations on the constraint matrix (linealities and prismatic rays) or by a more elaborate LP-based calculation of the ray matrix (conical rays). So, checking for compactness is generally no computational burden but can potentially facilitate an early exit.

9.1.2 *Open Simple Cones*

The opposite extreme is represented by (open) **simple cone** solution spaces. The 3D example sketched above, limited only by positive flux constraints before incorporating a fixed objective, is a simple illustration of an open simple cone polytope. The key property that defines a simple cone is that all the boundary hyperplanes are open but intersect at a single point, the apex of the cone. For the abovementioned 3D example, that is the coordinate origin, $f = (f_1, f_2, f_3) = (0, 0, 0)$.

For a metabolic model, the positive flux constraints are supplemented by the stoichiometric constraints. However, Eq. (1.1) shows that $f_i = 0$ also satisfies each stoichiometry constraint with an equality, which means that all stoichiometry constraint hyperplanes mutually intersect at the origin as well. So, in the absence of any further constraints such as flux bounds or an objective value, the solution space for a metabolic model without reversible reactions is still a simple cone with its apex at the origin. The additional boundary hyperplanes introduced by stoichiometry constraints mean that the SS only occupies a subset of the positive hyperquadrant formed by the coordinate axes, which can be pictured as a cone with top angles at the apex that are narrower than a right angle.

As argued in Chapter 4, reversible reactions correspond to linealities. After eliminating these, the remaining reduced SS is again a simple cone for metabolic models that contain reversible reactions but no objective.

The additional imposition of an objective value does not necessarily remove the simple cone behaviour. In the 3D example, an objective $f_1 = \phi$ would limit the SS to the (f_2, f_3) plane but it remains a 2D open cone.

In the simple cone, boundary facets are only constrained by intersections between constraint hyperplanes, and these ultimately form edges that extend to infinity but all intersect at the apex. These edges define extreme ray directions (see Chapter 4) and all interior points can be reached by a multiple of a convex combination of extreme rays.

Since the apex is common to all boundary hyperplanes and edges, it follows that for a simple cone the SS kernel is just a *single point*, namely the apex. Any other feasible point can be reached from it by the addition of a suitable multiple of a ray vector.

Such a single point SSK does not seem particularly interesting, particularly if it is merely the trivial point of zero flux at the origin.

However, even in this case there may be some relevant information encapsulated in the *shape* of the cone, for example, how 'sharp' it is and the shape of its cross-section.

This can be accommodated in the shape analysis framework of main chords and periphery points presented in the previous chapter, by extending the SSK from its minimal realisation as a single apex point, to the finite hypercone formed by tangent capping of the open hypercone at an *arbitrary finite cutoff* distance from the apex.

This extended version gives a compact subspace of the solution space that is still a valid SSK: it remains true that any feasible point can be reached by adding a multiple of a ray vector to a point within the SSK. Its size becomes arbitrary, determined by the choice made for the cutoff value, but its shape does reflect how the constraints of the metabolic model interact. For example, this would still reveal directions in flux space along which flux values are narrowly constrained compared to others where they can range more widely. If the cutoff value is chosen as a physically plausible maximum total flux, the resulting SSK should be correspondingly plausible.

If this approach is applied to the full flux space, it does not benefit from any available dimension reduction, and hence shape calculations, such as determining main chords, are likely to be computationally intractable for most metabolic models. For this reason, identification of a simple cone SS is not used to terminate the SSK calculation even though it is known to have a single point kernel. Instead, the RSS calculation discussed in Chapter 4, Section 2, is done first and usually results in a large dimension reduction. As a simple cone has no bounded facets, the concept of a progenitor facet does not apply and the calculation moves directly to tangent capping with the chosen arbitrary capping radius applied to all rays that remain in the RSS.

The described situation of a metabolic model without a defined objective, leading to a simple cone SS, most commonly occurs with genome scale models of multicellular organisms. In these cases, each cell type is likely to have a derived model, simplified to allow for gene expression patterns in the cell type and with an objective based on the physiological function of the specific tissue being modelled. Such specific models are more relevant to applications, and accordingly lead to SSK analysis of more interest, for example to bioengineering projects. From this perspective, the simple cone SS of a generic model is perhaps not worth pursuing further.

Where there is an objective that has been optimised, its value is unlikely to be zero and so it will usually rule out the flux space origin as a solution. That may nevertheless still give a simple cone SS at some stage of the SSK procedure, albeit with the cone apex at a different point. So, it is useful to have straightforward test for whether the SS is a simple cone.

The test can be made by checking if all constraint hyperplanes have a common intersection point. In terms of the SSK D-specification of Eq. (1.12), the constraint inequalities are converted to an equation

$$C \cdot F = V \tag{9.1}$$

Unlike in the original stoichiometry specification, the various transformations that give rise to the SSK usually produces a non-zero values vector V. The resulting non-homogenous equation is most easily solved by the use of the Moore–Penrose pseudoinverse C^+ as outlined in Eqs (5.6) and (5.7). This means that the cone apex point is given by

$$Apex = C^+ \cdot V \tag{9.2}$$

provided that the vector *Apex* also satisfies the equation

$$C \cdot Apex = V \tag{9.3}$$

If Eq. (9.3) is satisfied, the RSS is a simple cone and Eq. (9.2) gives the flux vector that constitutes the minimal, single point SSK at the apex; if not, the RSS is classified as a facetted cone.

9.1.3 *A Single Point SSK*

It can occasionally happen that a metabolic model has a fully determined flux value, either as originally specified or after SS reduction by removal of linealities and prismatic rays. This is a special case of a compact SS, and unlike the case of the single point SSK for a simple cone, shape analysis is not applicable. In this case, the D-specification of the SSK as in Eq. (1.12) contains no constraints and no SSK basis vectors and merely identifies the single allowable flux as the origin vector O in that equation.

9.1.4 *The Facetted Cone SSK*

The most general, and most common, SS type is one that (like a cone) is open with multiple ray directions, but for which there are both open and closed boundary facets, and no single common intersection point. This is designated as a **facetted cone**. The adjective is chosen to emphasise the presence of *bounded* facets, rather than its unbounded facets that are also present in a simple cone.

The concept is illustrated for a 3D case in Figure 1.4, albeit for the special case with just one ray direction, which is perhaps more accurately described as a facetted prism. Its generalisation to a facetted conical shape where the 10 unbounded sides geometrically diverge is easily visualised.

It is for the facetted cone that all the machinery to detect rays and bounded facets, capping and SSK shape and size analysis presented in this work are designed. Section 9.2 shows a variety of examples of this in realistic metabolic models.

9.2 Calculated SSKs for a Selection of Metabolic Models

This section introduces a number of metabolic models ordered approximately in the order of increasing complexity of the SSK calculations.

The results were obtained from the software implementation *SSKernel* based on this book, and which is available as open source *Mathematica* [2] code from the public domain software depositary GitHub.

9.2.1 *Escherichia coli Core model, e_coli_core [3]*

This is not a genome scale model, only including the core metabolism of *E. coli* and was designed as an educational example. It contains 263 constraints on 95 flux components. Reduction of the solution space, which includes application of the stoichiometry and a fixed objective value, identifies 48 flux components with fixed values and leaves only one degree of freedom. This turns out to be a single prismatic ray, so the RSS is a single point and this is also the final SSK. The actual flux values for this point is delivered as the origin point in the D-specification of the SSK, and also redundantly as a single peripheral point. Inspection of the ray vector reveals that the single degree of freedom is an equal mixture of only two out of the 95 flux components.

9.2.2 *Homo sapiens Red Blood Cell, iAB_RBC_283 [4]*

This model starts with 1281 constraints on 469 flux variables. A total of 271 fixed values are found during the reduction stage, which also removes three prismatic rays and a lineality ray pair. This leaves a compact RSS in eight dimensions, and as for the previous model this is also the final SSK. Two reactions specified as reversible in the original model are assigned a fixed direction by the combined constraints.

Shape analysis finds that the 60 peripheral points found give a radial coverage of 80%, and an estimated mean SSK diameter of 3.14 flux units. The graphical display of chords and diameters in Figure 9.1 gives a more detailed depiction of the shape.

For a more detailed description of how the graphics are interpreted, the reader is referred to Chapter 7, Section 4, where exactly the same red blood cell (RBC) model was used for illustration. Comparison of Figure 9.1 with Figure 7.8 shows the slight variations between separate runs of the SSK analysis on the same input data, as a result of random sampling effects.

9.2.3 *Helicobacter pylori, iIT341 [5]*

This model starts off with 1594 constraints on 554 flux variables, and reduces to an RSS of only 14 constraints on eight variables after removing

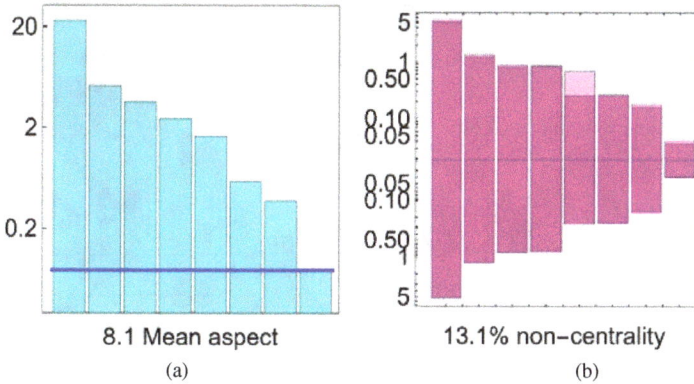

8.1 Mean aspect   13.1% non–centrality

(a)   (b)

Figure 9.1   The SSK for the human red blood cell model iAB_RBC_283: a compact RSS, no capping needed; showing (a) mutually orthogonal main chord lengths and (b) centered diameters along the chord directions. Refer to Figure 7.8 for the detailed description of the figure interpretation.

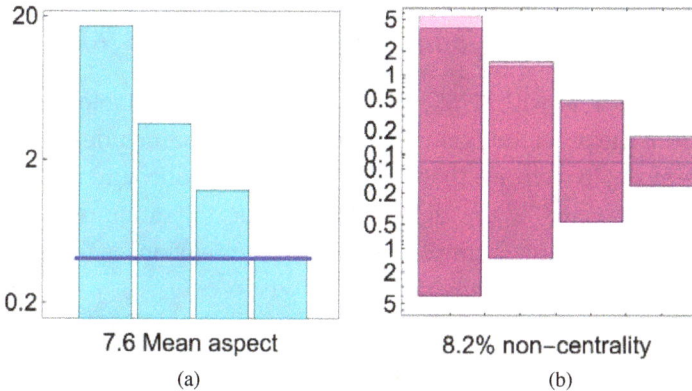

7.6 Mean aspect   8.2% non–centrality

(a)   (b)

Figure 9.2   The SSK for the Helicobacter model iIT341: Only coincidence capping applied to a facetted cone RSS; showing (a) mutually orthogonal main chord lengths and (b) centered diameters along the chord directions. Refer to Figure 7.8 for the detailed description of the figure interpretation.

350 fixed fluxes and six rays. The RSS is a facetted cone with only a single base level bounded facet (BFBF) and two remaining rays; these are eliminated by coincidence capping to give a six-dimensional compact SSK without the need for tangent capping. Flattening of two dimensions reduces the maximal aspect ratio below 50 and removes a mean chord and diameter conflict, giving a mean diameter of 4.03 and the values displayed in Figure 9.2.

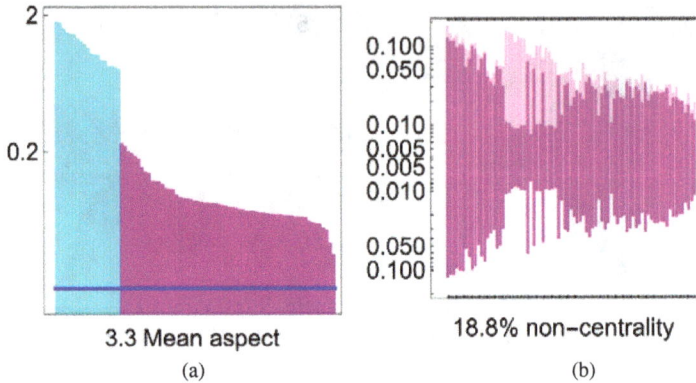

Figure 9.3 The SSK for the blood platelet model iAT_PLT_636: an open cone RSS, with 87 dimensions capped at a default radius of 0.3; showing (a) mutually orthogonal main chord lengths and (b) centered diameters along the chord directions. The colour change in (a) indicates the change from LP-calculated chord lengths to their diameter-based approximation.

9.2.4 *Homo sapiens Blood Platelet, iAT_PLT_636 [6]*

This model starts with 2755 constraints on 1008 variables, but is different from the previous examples in that it does not include an objective. It is therefore no surprise that the SS is a simple cone, and that no fixed fluxes are detected at the default fixed value tolerance of 0.002. The resulting RSS remains a simple cone in 242 dimensions after removal of 46 linealities and one prismatic ray.

Closer inspection, however, reveals that the inscribed hypersphere diameter is 0.01, much smaller than the chosen default capping radius of 0.3 flux units. So, the inscribed hypersphere is limited not by the capping cutoff, but rather the cone must be very thin in at least one dimension. Such thin dimensions can be eliminated early on by relaxing the fixed value tolerance to 0.05. At this setting, 325 fluxes are detected as fixed, giving the RSS as a simple cone in 99 dimensions.

Capping this at the default radius, it turns out that the inscribed diameter remains below the nominal fixed tolerance, so that further flattening is indicated. Flattening out all directions with chord lengths below 0.07 leaves the SSK as a default capped, simple cone in 87 dimensions, with calculated chords and diameters as shown in Figure 9.3.

It is clear at a glance that the SSK of the blood platelet model is vastly different from that of the RBC. Due to the absence of an optimised objective, it has a high dimensionality shown by the large number of main chords. The relatively low mean aspect ratio suggests a fairly isotropic cone 'cross-section', although the low value of the inscribed diameter (shown as a blue line in Figure 9.3(a)) compared to the smallest chord length, indicates that some thin directions have not yet been successfully detected. The mean diameter at 0.32 seems compatible with the capping radius, but still shows a conflict with the chord-based range estimate, which is another indication for further flattening. Unfortunately, in flattening out 12 dimensions to arrive at this picture, sufficient errors have accumulated that further flattening turns out to be unsuccessful.

It is seen that the shape analysis for an open cone SSK, while possible in principle, contains some ambiguity and does not yield as much simplification or information, and so is of lesser interest than for a typical facetted cone SSK.

9.2.5 *Methanosarcina barkeri, iAF692 [7]*

This model proceeds through all stages including tangent capping, but is still rather simple with regard to finding FBFs. It starts from 2009 constraints on 690 fluxes; separating off 463 fixed fluxes yields a facetted cone RSS with just nine constraints on six variables. Only three rays remain, and the resulting ray matrix shows that all constraint hyperplanes belong to the essentials set E (see Chapter 6, Section 6.6.1). This means that there is only a single BFBF, the tree search becomes trivial and tangent capping proceeds without any of the ambiguity that might arise from random searches. Flattening of three dimensions brings the inscribed hypersphere diameter in agreement with the smallest chord length without involving significant flattening error. This produces the 3D SSK with an estimated mean diameter of 0.83 and depicted in Figure 9.4.

9.2.6 *Clostridium ljungdahlii, iHN637 [8]*

This is a somewhat larger model that starts from 2269 constraints on 785 fluxes, of which 547 are determined to have fixed values. Its facetted cone RSS has 16 constraints on nine variables and just a single ray, leading again to a single BFBF and straightforward tangent capping. The capped SSK shows a clear separation of the chord lengths into four larger ones with single-digit values, and five small values of 0.025 and below. Flattening out the latter gives the four-dimensional SSK in Figure 9.5.

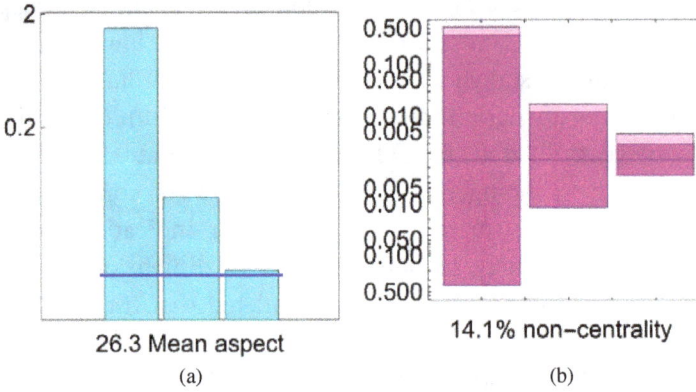

Figure 9.4 The SSK for the *M. barkeri* model iAF692: a facetted cone RSS, capped to preserve all FBFs and flattened from six to three dimensions; showing (a) mutually orthogonal main chord lengths and (b) centered diameters along the chord directions.

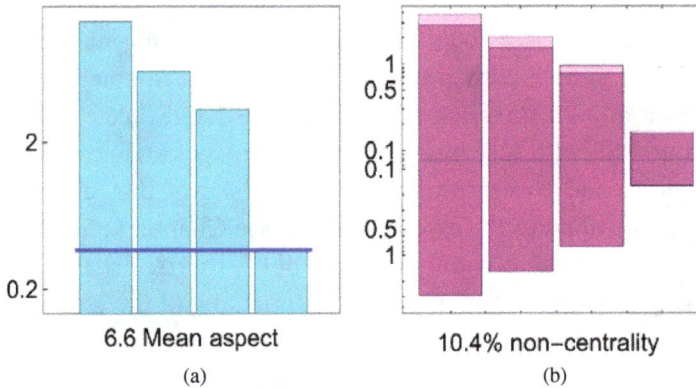

Figure 9.5 The SSK for the *Clostridium* model iHN637: a facetted cone RSS, capped to preserve all FBFs and flattened from nine to four dimensions; showing (a) mutually orthogonal main chord lengths and (b) centered diameters along the chord directions.

Comparing to the previous example, *M. barkeri*, the SSK calculations proceed similarly, but with a much smaller mean aspect ratio the Clostridium SSK is less anisotropic and its mean diameter estimate of 3.4 indicates a much larger hypervolume in flux space.

9.2.7 *Mycobacterium tuberculosis H37Rv, iNJ661 [9] and iEK1008 [10]*

This example allows the comparison of two different models for exactly the same bacterium: the original model iNJ661 set up in 2007, and an

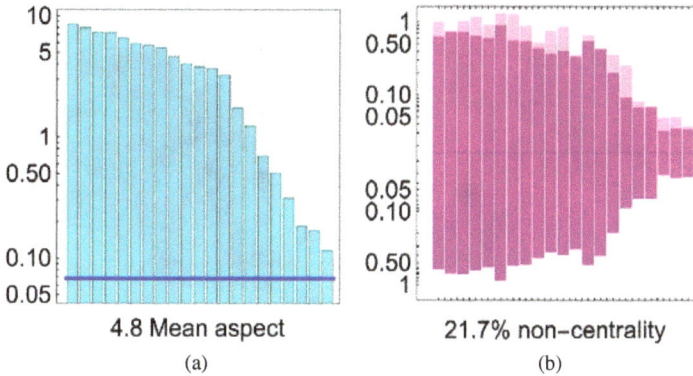

Figure 9.6 The SSK for the *Mycobacterium* model iNJ661: a facetted cone RSS, tangent capped to preserve all FBFs; showing (a) mutually orthogonal main chord lengths and (b) centered diameters along the chord directions.

extended version iEK1008, updated by the same research group in 2018.

The iNJ661 model starts from 2876 constraints on 1025 flux variables. Based on aspect ratios, the fixed value tolerance is set at 0.05 and the reduction stage fixes 709 fluxes and yields an RSS with 28 constraints on 21 variables and seven remaining ray vectors. The BFBFs are found exactly, as there are only two of them, and yield a progenitor facet at level 6 of the facet tree. This has no orthogonal rays, so seven tangent capping constraints are added to give a final SSK with 36 constraints on 21 variables as shown in Figure 9.6. The mean SSK diameter is 3.1 flux units for this SSK.

The iEK1008 model is considerably extended to 3451 constraints on 1226 fluxes, but at the same *fixtol* value of 0.05 there are 925 fixed fluxes leading to an RSS with even smaller dimensions of 15 constraints on 11 variables. It only has a single BFBF that is its own progenitor, and a single ray, again not orthogonal. The final SSK obtained after tangent capping the ray has a 16 × 11 constraints matrix and appears in Figure 9.7. It has a mean diameter of 9.6 flux units.

The progression of the SSK calculation is remarkably similar for both models, with the larger number of fluxes in the updated model more than compensated for by the fact that there are also more fixed values. The larger model in fact has only about half the number of free variables in the final SSK. Also, the number of ray vectors that represent unbounded degrees of freedom in the full SS is reduced from 22 in the original model to seven in the updated version.

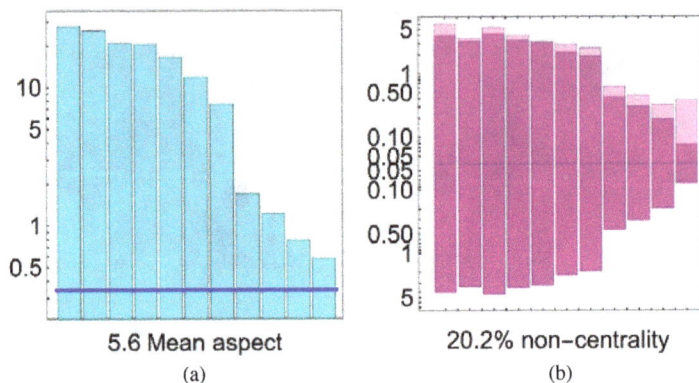

Figure 9.7 The SSK for the *Mycobacterium* model iEK1008: a facetted cone RSS, with one ray capped to preserve the single BFBF; showing (a) mutually orthogonal main chord lengths and (b) centered diameters along the chord directions.

Regarding shape, a notable feature in Figure 9.6 is the separation of chord lengths into two ranges with a steeper decline for the lower range. The same feature is visible in Figure 9.7 even if for fewer dimensions. Both SSKs have quite similar mean aspect ratios. The estimated mean diameter of the refined model is three times larger than for the original.

The noted similarities between the two sets of results may reflect the fact that they represent inherent features of the bacterium rather than just the models. A more mundane explanation would ascribe the similarities to the fact that both models have very similar, although not identical, objectives that are maximised. The objective of the larger model contains contributions from additional fluxes, but for corresponding fluxes the contributions are almost the same, and the optimised objective values of 0.0525 and 0.0581, respectively, are almost the same. In fact, these two explanations amount to much the same thing; the objective in an FBA calculation summarises the effects of metabolic regulation and so characterises the intrinsic metabolic state of a cell rather than the model that was set up.

9.2.8 *Salinispora: Core Model iCore_BMC, Arenicola iCCSA643_BMC and Pacifica iCCSP114_BMC [11]*

This example is a comparison that contrasts with that of section 9.2.7 - here, the three models from the same study are compared: one for

each of two different species of the *Salinispora* genus of bacteria, and a core model meant to model the metabolic capabilities common to multiple *Salinispora* species and strains.

Starting with the core model, it has a considerably sized constraint matrix of dimensions 2763×876. The 876 flux count is larger than some of the microbe examples discussed previously. Nevertheless, after fixing 458 of these the RSS reduces to a moderate 61×54 constraints matrix. The facet tree is small enough to be exhaustively searched and yields just a single BFBF at level 47. Coincidence capping to this effective progenitor eliminates all 52 rays that remain and gives a compact SSK as 14 constraints in seven dimensions without tangent capping. Of these five turn out to be thin directions that are flattened out to avoid excessive aspect ratios. The final, almost trivial SSK is as shown in Figure 9.8 and has a mean diameter of 1.2 flux units.

Compared to the smaller models shown before, this SSK is very constrained: it has only two significant degrees of freedom, and even for these the range of variation is only two units along one direction and 0.4 units along the second direction. That is quite plausible for a core model, showing that the core metabolism is quite closely dictated and does not leave much room for being externally manipulated.

The *Arenicola* model behaves rather differently. It starts from a 3314×1112 constraint matrix, fixes 489 fluxes and delivers a 133×110 constraint matrix to define the RSS. The 110 flux dimensions requires a randomised greedy search to find a single BFBF at tree level 82, that is

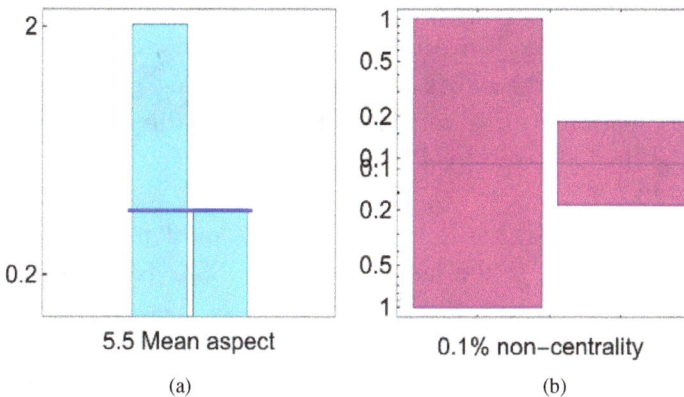

(a)   (b)

Figure 9.8   The *Salinospora* core model iCore_BMC: a compact SSK after coincidence capping to 7 dimensions and flattened to two; showing (a) mutually orthogonal main chord lengths and (b) centered diameters along the chord directions.

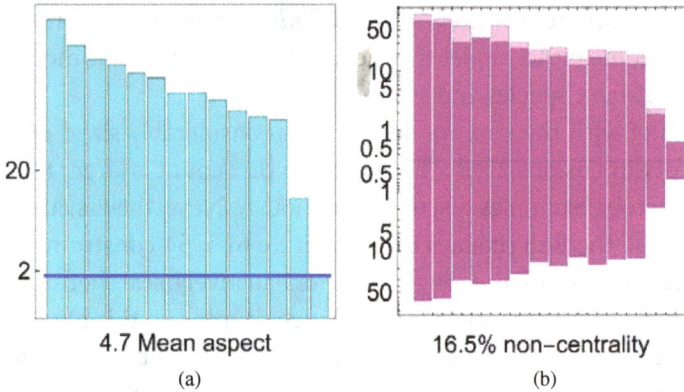

Figure 9.9 The *Salinospora arenicola* model iCCSA643_BMC: a facetted cone, capped and flattened from 30 to 14 dimensions; showing (a) mutually orthogonal main chord lengths and (b) centered diameters along the chord directions.

traced back to a progenitor at level 80. Coincidence capping to the progenitor eliminates 82 rays and reduces the dimensions to 30, low enough that an exhaustive search becomes possible and yields two BFBFs. The final tangent capping disposes of the remaining seven rays resulting in 61 constraints on 30 variables. Finally, 16 dimensions with chord lengths smaller than 0.275 are flattened out giving the final SSK with 49 constraints on 14 variables as shown in Figure 9.9.

The *Salinispora pacifica* model follows a similar path. From a somewhat larger initial 4019 × 1363 constraint matrix, it reduces to an RSS with a 207 × 182 constraint matrix. This time the randomised greedy search finds 11 BFBFs that trace back to a progenitor at level 136. Coincidence capping then leaves 25 rays; a random greedy search is still needed and finds 12 BFBFs, and tangent capping to preserve these gives 95 constraints on 46 variables. Flattening 17 dimensions below 0.7 yields the final SSK specified by a 91 × 29 constraint matrix shown in Figure 9.10.

For both species, the feature that differs most markedly from all previous examples is the large size of the SSK. For *Arenicola*, the maximal chord length is 623 flux units and the mean diameter is 137, whereas for *Pacifica* the maximal chord length is a similar 532 but the mean diameter is even larger than for *Arenicola* at 167. Their respective mean aspect ratios of 4.7 and 5.2 are quite low, indicating a relatively isotropic shape and this is further illustrated by the fairly uniform decline in chord lengths and diameter values shown in the figures. There

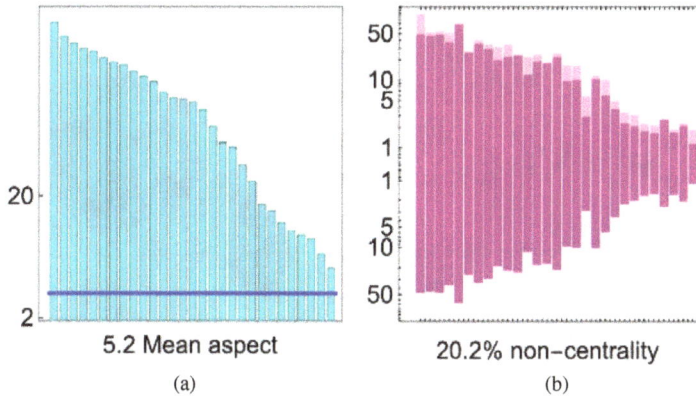

Figure 9.10 The *Salinospora pacifica* model iCCSP114_BMC: a facetted cone, capped and flattened from 46 to 29 dimensions; showing (a) mutually orthogonal main chord lengths and (b) centered diameters along the chord directions.

is nevertheless a considerable difference between the SSKs of the two species, with *Pacifica* having more than twice the number of degrees of freedom and a steeper decline in chord lengths.

Taken together with the restricted range of the core metabolism of *Salinispora* species as noted earlier, it appears that most of the wide overall variation that these two models allow is associated with metabolites and pathways that are outside the core. It is intriguing to notice that the very reason *Salinispora* species receive research interest is that they are known to produce specialised metabolites such as antibiotics, anticancer agents and unique pigmentation [11]. Perhaps, the large metabolic variability indicated by the large SSK is an adaptation of the organism to a varied marine environment and can be exploited for bioengineering, but a more detailed examination of such questions is beyond the scope of this study.

9.2.9 *Pseudomonas putida iJN746 [12] and iJN1463 [13]*

This is another case where an older model was later expanded into a much larger version. The 2008 model iJN746 turns out to be somewhat problematical for the SSK analysis despite a moderate 1054 flux count. With the default fixed value tolerance of 0.002 the RSS has 115 dimensions and this remains the same after capping. Its main chord lengths vary from <0.1 to >1000, giving a very large aspect ratio >10^4 that demands flattening. However, the decline of values is quite gradual for the lower half of lengths giving no obvious dividing line below which to flatten.

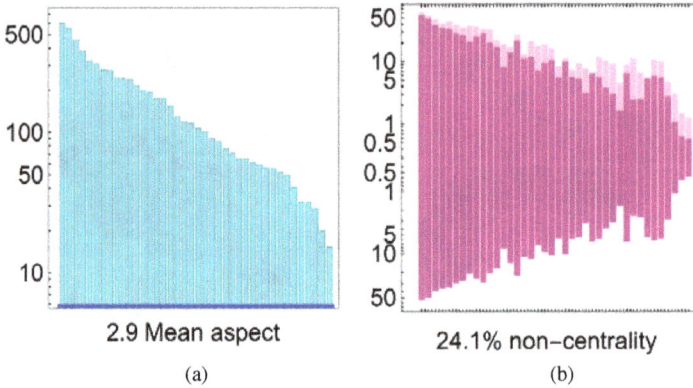

Figure 9.11 The *Pseudomonas putida* model iJN746: a facetted cone, tangent capped and flattened to 40 dimensions; showing (a) mutually orthogonal main chord lengths and (b) centered diameters along the chord directions.

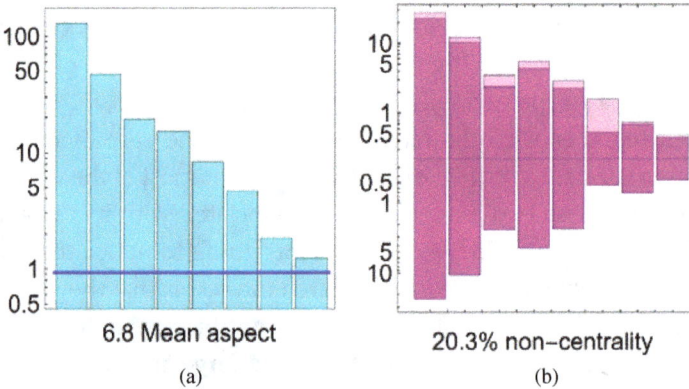

Figure 9.12 The *Pseudomonas putida* model iJN1463: a facetted cone RSS, becoming bounded in eight dimensions after coincidence capping; showing (a) mutually orthogonal main chord lengths and (b) centered diameters along the chord directions.

This is a typical example of a model that requires preemptively setting *fixval* to 0.45, in order to limit the aspect ratio. This yields 568 fixed fluxes and the RSS reduces to 86 constraints on 68 flux variables. Coincidence capping to a level 25 progenitor found by a randomised greedy search, leaves nine rays that are tangent capped to yield a 71 × 43 constraints matrix. The maximal aspect ratio remains high, but now there are three distinctly separated thin directions and flattening these away reduces the maximal aspect ratio below 40. The estimated diameter of 174 confirms the chosen fixed value tolerance as a consistent choice for negligible variation, and so does the lower limit as supplied by the inscribed hypersphere diameter of 5.9. The result is shown in Figure 9.11.

Moving on to the 2020 update as model iJN1463, it is a major expansion with 2927 flux variables (a threefold increase) and subject to 8008 constraints. The objectives for the two models are similar, but not the same, and the optimised values differ considerably. Using the same value *fixtol* = 0.45, a total of 2073 fixed fluxes are isolated for this model, giving a 56 × 51 constraint matrix for its RSS. The smaller space dimension allows an exhaustive facet tree search and finds a single BFBF as its own progenitor at level 43. Coincidence capping is again possible and yields a closed SSK without needing tangent capping, at a strikingly low 13 constraints on the remaining eight flux variables. The resulting SSK with a mean diameter estimate of 45.5 is shown in Figure 9.12.

Comparing the results, the most striking aspect is that although the revised model has about three times the total number of flux variables, its SSK has fewer degrees of freedom by a factor of 5 than the original model.

This is slightly offset by an increase in the ray space dimensions from 49 for iJN746 to 56 for iJN1463. In addition, the range of variation indicated by the estimated mean SSK diameter of the reduced set of flux variables is smaller by a factor of almost 4. So, despite its increased size, the revised model pins down the metabolism to a larger extent than for the original version.

Regarding SSK shape, both versions display a smooth decline in chord lengths and have quite low mean aspect ratios.

By and large, these observations are similar to those for the *Mycobacterium* example. In both cases, the additional reactions and metabolites in the elaborated model reduce the number of degrees of freedom of the SSK and their range of variation, while retaining some of the shape features. This is not a matter of model size, as shown by the opposite tendency when comparing the smaller core *Salinispora* model with both of the extended versions for individual species that have much larger SSKs. Rather, it seems that the revised models have increased the complexity of the metabolic network, perhaps introducing additional cross-linking that increases restrictions on flux values. Even so, both the original and elaborated versions are consistent in predicting that the *Pseudomonas* SSK has a much larger diameter than that of *Mycobacterium*, suggesting this as an intrinsic characteristic of the organism rather than the model.

A plausible interpretation is that when the SSK is large in terms of dimensions (degrees of freedom), diameter (size) and number of ray directions, one has to be cautious about ascribing this to a fundamental feature of the organism and its metabolism. It could also merely be a consequence of an incompletely known metabolic network, that is, an inadequate model. That particularly applies when there are many rays, as these reflect an infinite range that intrinsically conflicts with physical reality.

9.2.10 *Geobacter sulfurreducens [14–15] and Metallireducens iAF987 [16]*

The *Sulfurreducens* model is the one first introduced in Chapter 1, and used subsequently for illustrative purposes. It is a medium-sized model with 2586 constraints on 940 flux variables. A first run establishes that the maximal SSK chord length value is large at about 460, so the fixed value tolerance is set at 0.5 to give 481 fixed fluxes and this yields a facetcone RSS with a 123 × 101 constraints matrix. A randomised greedy search finds multiple BFBFs at four different levels ranging from 74 to 79, with a progenitor at level 54. That is a far more complex situation than encountered in the previous examples. The progenitor allows coincidence capping to give 69 constraints on 43 variables, but with an extensive facet tree not amenable to exhaustive search. The greedy search done instead is stopped when it reaches 500 BFBFs, and these are used as a sample on which to base tangent capping. Flattening out four directions thinner than 0.05 yields a final 99 × 43 SSK constraints matrix. The shape characterisation is illustrated in Figure 9.13.

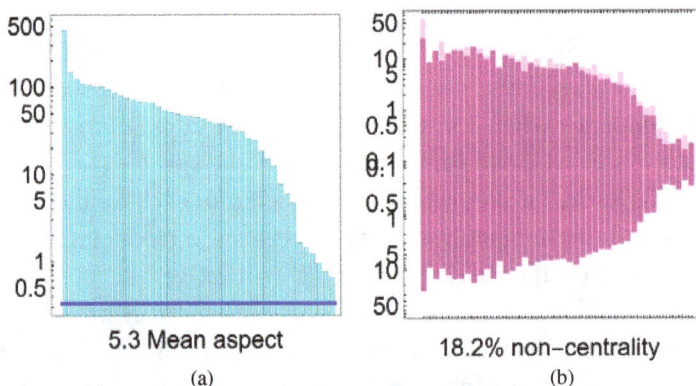

Figure 9.13   The *Geobacter sulfurreducens* model: a facetcone SSK in 43 dimensions; showing (a) mutually orthogonal main chord lengths and (b) centered diameters along the chord directions.

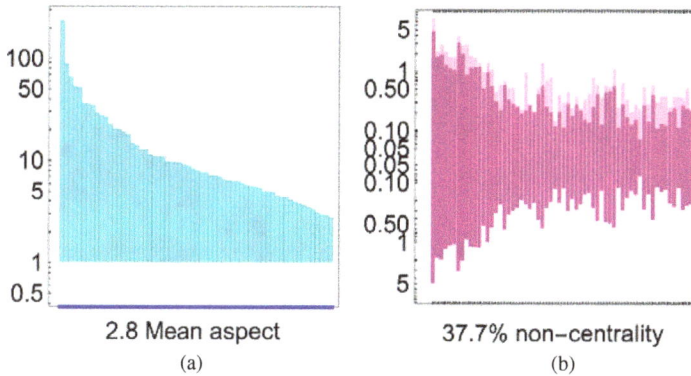

Figure 9.14 The *Geobacter metallireducens* model iAF987: a facetcone SSK in 62 dimensions; showing (a) mutually orthogonal main chord lengths and (b) centered diameters along the chord directions.

It is seen that there is a single long chord of length 455 and several short ones giving the maximal aspect ratio of 698, but most chords are fairly uniform giving a moderate mean aspect ratio of 5.3 and an estimated SSK mean diameter of 45.5.

Model iAF987 is again a revised version, further elaborated to model the metabolism of the different *Geobacter metallireducens* species. Compared to the *Sulfurreducens* model, its flux count is increased to 1285, but even with a somewhat stricter fixed value tolerance of 0.2, the number of fluxes identified as fixed is almost double at 792.

This leads to a 72 × 63 RSS constraints matrix, allowing exhaustive search to find four BFBFs with their progenitor at level 6. There is no orthogonal ray, so the calculation proceeds to tangent capping of the eight progenitor rays. One essentially zero chord length (a nominal value 0.0004) remains and is flattened out to give the final SSK with 81 constraints on 62 flux variables as pictured in Figure 9.14.

The *Metallireducens* SSK has a noticeably higher dimension count, but its 13 ray space dimensions are far fewer than the 247 ray dimensions of the *Sulfurreducens* model.

Overall, the larger, more elaborate model appears to have narrowed down uncertainties in having more fixed values and fewer unbounded flux values. Although it has a larger SSK dimension count, the estimated diameter of 15.2 is considerably smaller than the 45.5 of the smaller model. The overall shape trend shown by the orthogonal chord lengths is quite similar for both cases, as is the respective mean aspect ratios.

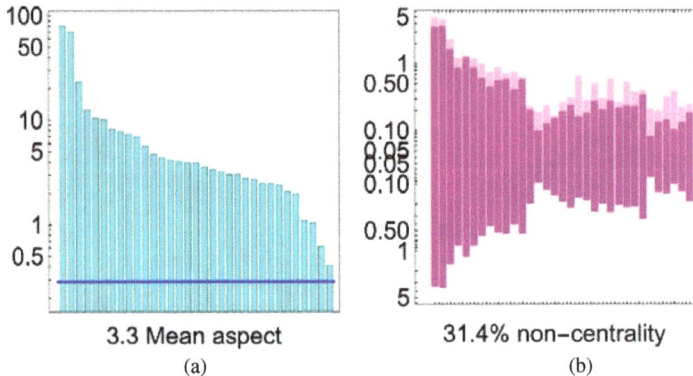

Figure 9.15 The SSK of *Escherichia coli K12* model iML1515: a facetcone SSK in 33 dimensions; showing (a) mutually orthogonal main chord lengths and (b) centered diameters along the chord directions.

9.2.11 *Escherichia coli K12, iML1515* [17]

This is a large model, starting from 7302 constraints on 2712 flux variables. Of these, 1945 are fixed within a tolerance 0.05, leading to a 49 × 39 constraints matrix for the RSS. Exhaustive search identifies two BFBFs at different levels, but descended from a level 4 progenitor. After both coincidence and tangent capping, followed by flattening two thin directions, the final SSK has 53 constraints on 33 flux variables. It is complemented by a 27-dimensional ray space. The graphical depiction of its shape is given in Figure 9.15.

As might be expected for such a standard reference organism with a well-characterised metabolic network, the analysis conforms to the pattern that was observed above for other extensive models. About two-thirds of all flux values are essentially fixed; the optimised objective leaves an SS kernel with a modest number of degrees of freedom counting in the lower double digits, leaving a similar number of degrees of freedom with undetermined upper limits, that is, rays. The SSK has quite a small mean aspect ratio of 3.2 and a similarly modest estimated mean diameter of 8.4.

9.2.12 *Saccharomyces cerevisiae, iMM904* [18]

The previous examples were mostly bacteria, so this example extends the application to the common yeast cell often used as a model organism for eukaryotic cells. The model has 1577 flux variables, of which 1024 turn out to be fixed giving an RSS specified by 35 constraints on 24

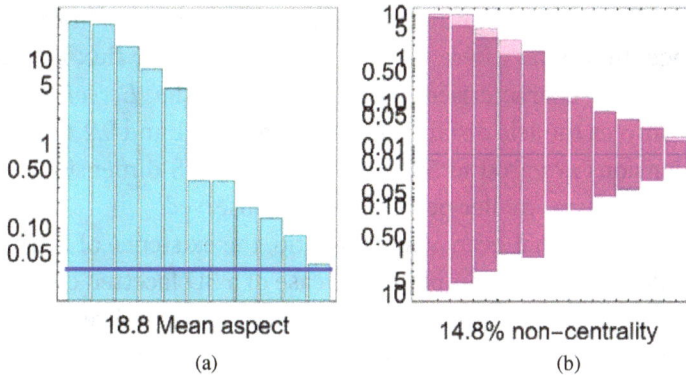

Figure 9.16 The SSK of *Saccharomyces cerevisiae*, model iMM904: a facetcone SSK in 11 dimensions; showing (a) mutually orthogonal main chord lengths and (b) centered diameters along the chord directions.

variables. Exhaustive search establishes a progenitor at level 10 of the facet tree, and the combination of coincidence and tangent capping followed by flattening out three directions that have thicknesses below 0.02 gives an SSK defined by a 24 × 11 constraints matrix as shown in Figure 9.16, supplemented by 31 ray dimensions.

The 11-dimensional SSK has an estimated diameter of 10.7 and a considerable mean aspect ratio of 19.2. It does not seem that the eukaryotic cell nature affected the analysis significantly and by and large the trends described for other well-characterised metabolic models apply here as well.

9.2.13 *Rattus norvegicus liver cell, iRatLiver [19]*

Despite not being excessively large, starting from 5213 constraints on 1882 variables, this model turns out to test the limits of the SSK calculation. In order to keep the RSS dimensions manageable, that is, below 200, the *fixtol* value has to be chosen around 0.4 to 0.5. Taking a value of 0.5 gives 774 fixed fluxes and an RSS with 140 dimensions. The progenitor is found at level 111, so that coincidence capping leaves only 29 flux variables.

A feature of this model that stands out compared to most of the models covered above is the large number of linealities (i.e., flux directions that remain completely unconcostrained) and rays, with the ray space dimension of 216 being only marginally smaller than the total flux dimension of 217 for the RSS before linealities and prismatic rays are removed. This seems significant, because rays and linealities indicate flux space directions where there is a lack of realistic flux limitations. For most well-characterised models, such as the *E. coli* K12 and Saccharomyces models discussed in sections 9.2.11 and 9.2.12, the total ray dimensions are around 50% to 60% of the RSS dimensions, and in some models even smaller percentages are observed.

Note, though, that even at a 100% ratio, the existence of FBFs would not be ruled out – for example, in the case of a 3D facetted cone similar to Figure 1.4, but with diverging sides, there would be three ray dimensions that are equal to the SS dimensions, but even so the figure shows several bounded facets.

Indeed, the coincidence capped rat liver model yields 15 BFBFs, but when these are used to determine tangent capping radii a second remarkable feature appears: the largest capping radius is 0.33, which is well below the value of 0.5 chosen for the fixed tolerance. This appears somewhat contradictory: having taken flux variations up to a value 0.5 as essentially constant, it does not make sense to consider variations of lesser magnitude over the scope of the SSK to be significant. So, for consistency, the SSK is dealt with as essentially a simple cone, but rather than a point apex it has a small flux hypervolume at its base. To implement this, the small calculated capping radii are all replaced by the larger user-chosen default capping radius. In this case, the value is taken also as 0.5, large enough that no FBF will be intersected by a capping hyperplane but small enough to be consistent with the fixed tolerance.

Proceeding with calculating the 29 orthogonal chord lengths, several of these are below 0.1 and flattening them out gives the final SSK as bounded by 62 constraint hyperplanes in 21 dimensions, and illustrated in Figure 9.17.

As the figure shows, the SSK has a low mean aspect ratio of 2.0, and the small estimated mean diameter of 0.6 merely reflects the default capping radius used.

To summarise, the rat liver model combines a kernel that is small in absolute size as well as dimensionality, with a large number of unconstrained degrees of freedom (rays).

The reliability of this conclusion may seem compromised by the apparently arbitrary choice that was made for the fixed tolerance. In other cases, where a similarly large value was chosen, such as for the *Pseudomonas* and *Geobacter* models, this could be retrospectively justified as being insignificant compared to the maximal chord lengths that

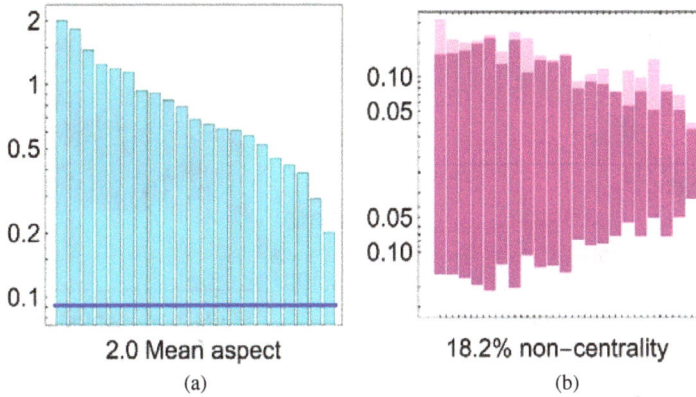

Figure 9.17 The SSK of the *Rattus norvegicus* liver cell model as a facetted cone SSK in 21 dimensions; showing (a) mutually orthogonal main chord lengths and (b) centered diameters along the chord directions.

were found. But here, the maximal chord length of the final SSK is only 2 flux units, not convincingly large compared to the 0.5 tolerance value.

An alternative approach is possible to test this. Leaving the fixed tolerance at the default value of 0.05 for a model of this size, only 84 fluxes are classified as fixed and this gives an RSS with 399 flux dimensions. This value is too large to make an FBF search practical, which also rules out dimension reduction by coincidence capping. However, it is still possible to perform tangent capping with a default radius of 0.5 for all 399 degrees of freedom, essentially treating it as a simple cone SSK. The chords can still be calculated, albeit using the diameter approximation for all but the first few chords.

The resulting shape characterisation appears as in Figure 9.18. As is expected, there is a large drop by an order of magnitude, from the first 10 LP calculated chords to the values estimated by diameters. A second drop is just discernible after chord no 16, beyond which almost all diameters fall in a range between 0.01 and 0.05. Even allowing for these to be order of magnitude underestimates of the chord lengths, essentially the same picture emerges as before: there are only 16 degrees of freedom with a modest range, and all others fall below 0.5. Also, a consistent lower limit of 0.36 is obtained for the mean SSK diameter, although neither the upper limit nor a value estimate is available because the circumscribed simplex calculation is also intractable for the large dimension count.

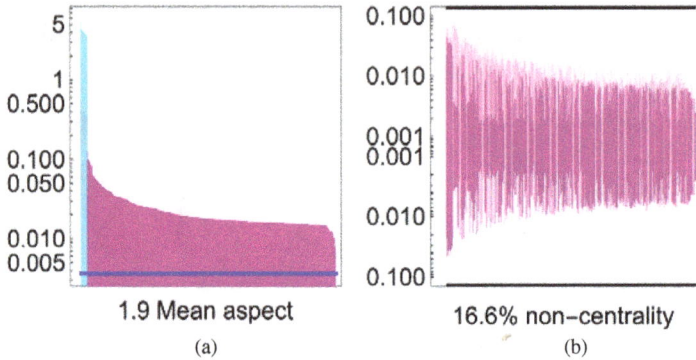

Figure 9.18   The SSK of the *Rattus norvegicus* liver cell model as a simple cone SSK in 399 dimensions; showing (a) mutually orthogonal main chord lengths (cyan) and diameters (magenta) and (b) centered diameters along the chord directions.

However, there is a considerable price to pay for the slightly more consistent procedure: not only is the SSK description far more cumbersome and less detailed, but it requires several hours of computing time, compared to the 6 min needed to do the calculation as first described with an adjusted fixed tolerance.

9.2.14 *Homo sapiens, RECON1 [20]*

To conclude the case studies, the RECON1 model of human metabolism allows investigating the applicability of the SSK analysis on a rather large model with 10,249 constraints on 3741 flux variables. This model presents a further obstacle to the analysis, in that although it does have a nominal objective, the objective turns out to have a zero value as its FBA optimum. This means that the solution space is a simple cone with its apex at the origin in flux space, that is, a trivial zero flux, and this single point is also the SSK.

That does not seem a particularly interesting or informative result, but is perhaps to be expected for a generic model that encompasses all cell types and environments in a multicellular organism. It is only when the model is specialised, for example, to a particular tissue, where the regulation of metabolism is geared to perform a particular physiological function, that the objective function that represents this regulation can be expected to put meaningful bounds to the FBA solution space.

As described in Section 9.1.2, one can nevertheless proceed with shape analysis of the cone using tangent capping of all rays with a default

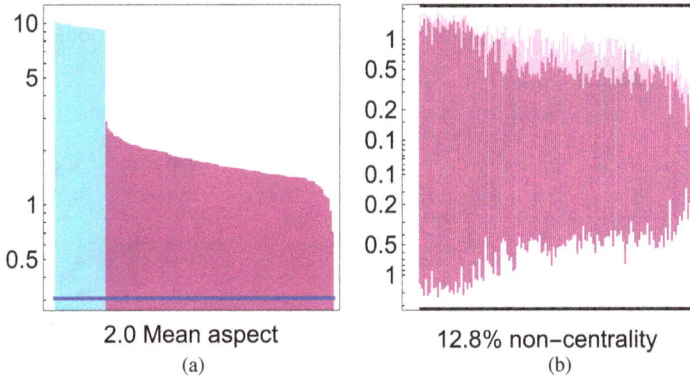

Figure 9.19 The SSK of the *Home sapiens* model RECON1 as a simple cone SSK in 137 dimensions; showing (a) 25 mutually orthogonal main chord lengths (cyan) and 112 diameters (magenta) and (b) centered diameters along the chord directions.

capping radius chosen as 1.0. For the RECON1 model, using the value *fixtol* = 0.05, which is the default for a model of this size, there are 1394 fixed fluxes that leaves the RSS as a simple cone in 474 dimensions. This is still too high for shape analysis to be practical. Increasing the fixed tolerance to the value of 0.3 projects out 1974 fixed fluxes and gives an RSS in a more amenable 152 dimensions.

The maximal chord calculation for this identifies a further 15 dimensions in which the capped kernel thicknesses are negligible (below 0.00002) and which are flattened out to deliver an SSK in 137 flux dimensions.

The results are illustrated in Figure 9.19. The estimated mean SSK diameter is 7.1, well justifying the chosen fixed tolerance value. The figure shows a fourfold drop from the LP calculated chords to diameter estimates, suggesting that in fact the vast majority of the maximal chords are at values above 5. This is in contrast to the case of the rat liver model in section 9.2.13 that looks superficially similar, but where the different scale indicated a much more constrained SSK cone.

9.3 Discussion and Future Perspectives

For an overview of the case study results, they are summarised in Table 9.1.

Table 9.1: Summary of calculations for example models, proceeding from the initial full flux space (FFS) to the reduced solution space (RSS) to the solution space kernel (SSK).

	Flux Dimension			Count		Shape		
	FFS	RSS	SSK	Fixed	Rays	Diameter mean	Aspect mean	Aspect max
E. Coli Core	95	0	0	48	1	0	—	—
Red Blood Cell	469	8	8	271	4	3.1	8.1	290
H. Pylori	554	8	4	350	6	4.0	7.6	29
Blood Platelet	1008	99	87	325	145	0.3	3.3	50
M. Barkerii	690	6	3	463	12	0.8	26.3	135
Clostridium	785	9	4	547	3	3.4	6.6	36
Mycobact iNJ661	1025	21	21	709	22	3.1	4.9	78
iEK1008	1226	11	11	925	7	9.6	5.6	48
Salinispora Core	876	54	2	458	64	1.2	5.5	6
S. Arenicola	1112	110	14	489	104	137	4.7	341
S. Pacifica	1363	182	29	508	183	167	5.2	104
Pseudom iJN746	1054	68	40	568	49	174	2.9	40
Pseudom iJN1463	2927	51	8	2073	56	45.5	6.8	105
Geobacter Sulf.	940	101	43	481	247	45.5	5.3	698
Geob. Metallired.	1285	63	62	792	13	15.2	2.8	83
E Coli iML1515	2712	39	33	1945	27	8.4	3.2	162
Saccaromyces	1577	24	11	1024	31	10.7	19.2	769
Rat liver	1882	140	21	774	216	0.6	2.0	16
H. Sap. RECON1	3741	152	137	1974	288	7.1	2.0	15

A number of lessons about SSKs and their calculation can be learned, at least tentatively, from the examples in this chapter. The first is that the kernel dimension is smaller than that of the original model flux dimension by around two orders of magnitude. Confining the FBA solutions to the null space of the stoichiometry matrix accounts for some of this reduction, but two other major contributors are the following:

- Separating off flux values that remain fixed throughout the SS. This usually amounts to a large fraction, 40%–70% of all fluxes. Most of this happens at the start using the Hop, Skip and Jump algorithm based on the tolerance value taken to define a fixed value, that is, the variable *fixtol*. These are supplemented in the final stages by flattening of SSK dimensions considered negligible, as judged by aspect ratios of the capped kernel. The divide between the a priori and a posteriori dimension reductions is largely controlled by the value that is chosen for *fixtol*.

- Elimination of ray directions can also give large dimension reductions. Most of this usually happens in the early stages of eliminating linealities and prismatic rays. A smaller contribution can come from coincidence capping to the progenitor hyperplane, but that is not feasible in all models.

As a result of these reductions, the size of the original model is not always the best indication of how tractable the SSK computation is. Instead, the dimension count of the RSS, where the most straightforward of the reductions have been done, is a better indicator of whether all (or at least a representative sample) of the bounded facets can be found.

There is quite a large variation between different models in the course of the calculation and whether approximations are needed. Remarkably, though, the SSK dimensions can fairly consistently be kept to a single- or double-digit number. Higher numbers are usually associated with simple cone solution spaces, where the objective is absent or trivial and so does not introduce substantial limitations. Even for the same model, different choices of calculation parameters can lead to slightly different SSK dimension counts.

Differences in both model size and how the calculation proceeds, mean that the computational effort required, ranges from a couple of seconds for simple cases to several hours of computing time for the more challenging ones.

The hypervolume occupied by the SSK can be compared between models by estimating the mean diameter of each. Disregarding simple cone cases where the diameter is at least partially determined by the chosen capping radius, the values as calculated for the examples presented from their intrinsic bounded facets vary widely between <1 and as high as 174. This implies that the metabolic range accessible to different cells can be quite different depending on the interplay between constraints, which is induced by the structure of a particular network.

The shapes of SSKs can be broadly described by their set of aspect ratios between the pairs of orthogonal chords. The mean aspect ratio is far less variable than the size, with typical values between 3 and 7. The maximal aspect ratio, indicating how the longest and shortest chords compare, is usually between 50 and 80, but in some cases can rise to several hundred because of one or two chords that are exceptionally long or short.

The question arises whether the observed variations are biologically significant. The first issue is whether it reflects real metabolic differences or merely the state of the metabolic model. Comparisons between different models for the same organism and between different organisms provided evidence for both model-related and real differences. From the case studies, it seems plausible to suggest that the size of the SSK may give an indication of the metabolic adaptability or degree of specialisation of the organism being modelled. However, much more detailed studies will obviously be needed to make firm conclusions.

To place this into context, it is useful to consider the construction of an FBA SSK as analogous to producing a microscope. In a somewhat similar way as a microscope is a tool to allow details of the physical structure of a cell to be discovered and studied, the SSK is a tool that allows the user to penetrate some of the vast complexity associated with conventional descriptions of the metabolic space. It puts the focus on some essential features and allows further study by reducing the solution space to a more manageable object.

The subject of this book is the design concepts and the actual construction of this new 'microscope'. In the course of testing its application on actual examples, some intriguing observations suggest possible areas for further study.

But just as the inventor of a physical microscope might point out a few structures that become visible through his lenses in order to attract the interest of the cell biologist, it is hoped that the discussion above might stimulate those with expert knowledge of the biochemistry to use this new tool to gain a better understanding of the metabolism of cells, perhaps leading to new opportunities for bioengineering.

References

1. Z. A. King, J. Lu, A. Dräger, P. Miller, S. Federowicz, J. A. Lerman, A. Ebrahim, B. O. Palsson, & N. E. Lewis, BiGG models: a platform for integrating, standardizing and sharing genome-scale models. *Nucleic Acids Research*, **44** (2016) D515–D522. https://doi.org/10.1093/nar/gkv1049.
2. Wolfram_Research & Inc., *Mathematica*, Version 12 (Champaign, Illinois: Wolfram Research, Inc., 2021).

3. J. D. Orth, B. Ø. Palsson, & R. M. T. Fleming, Reconstruction and use of microbial metabolic networks: the Core *Escherichia coli* Metabolic Model as an educational guide. *EcoSal Plus*, **4** (2010). https://doi.org/10.1128/ecosalplus.10.2.1.

4. A. Bordbar, N. Jamshidi, & B. Ø. Palsson, iAB-RBC-283: a proteomically derived knowledge-base of erythrocyte metabolism that can be used to simulate its physiological and patho-physiological states. *BMC Systems Biology*, **5** (2011). https://doi.org/10.1186/1752-0509-5-110.

5. I. Thiele, T. D. Vo, N. D. Price, & B. Ø. Palsson, Expanded metabolic reconstruction of Helicobacter pylori (iIT341 GSM/GPR): an in silico genome-scale characterization of single- and double-deletion mutants. *Journal of Bacteriology*, **187** (2005) 5818–5830. https://doi.org/10.1128/JB.187.16.5818-5830.2005.

6. A. Thomas, S. Rahmanian, A. Bordbar, B. Ø. Palsson, & N. Jamshidi, Network reconstruction of platelet metabolism identifies metabolic signature for aspirin resistance. *Scientific Reports*, **4** (2014). https://doi.org/10.1038/SREP03925.

7. A. M. Feist, J. Scholten, B. Ø. Palsson, F. Brockman, & T. Ideker, Modeling methanogenesis with a genome-scale metabolic reconstruction of Methanosarcina barkeri. *Molecular Systems Biology*, **2** (2006). https://doi.org/10.1038/msb4100046.

8. H. Nagarajan, M. Sahin, J. Nogales, H. Latif, D. R. Lovley, A. Ebrahim, & K. Zengler, Characterizing acetogenic metabolism using a genome-scale metabolic reconstruction of *Clostridium ljungdahlii*. *Microbial Cell Factories*, **12** (2013). https://doi.org/10.1186/1475-2859-12-118.

9. N. Jamshidi, & B. Ø. Palsson, Investigating the metabolic capabilities of *Mycobacterium tuberculosis* H37Rv using the in silico strain iNJ661 and proposing alternative drug targets. *BMC Systems Biology*, **1** (2007). https://doi.org/10.1186/1752-0509-1-26.

10. E. S. Kavvas, Y. Seif, J. T. Yurkovich, C. Norsigian, S. Poudel, W. W. Greenwald, S. Ghatak, B. Ø. Palsson, & J. M. Monk, Updated and standardized genome-scale reconstruction of *Mycobacterium tuberculosis* H37Rv, iEK1011, simulates flux states indicative of physiological conditions. *BMC Systems Biology*, **12** (2018). https://doi.org/10.1186/S12918-018-0557-Y.

11. C. A. Contador, V. Rodríguez, B. A. Andrews, & J. A. Asenjo, Use of genome-scale models to get new insights into the marine actinomycete genus Salinispora. *BMC Systems Biology*, **13** (2019) 1–14. https://doi.org/10.1186/S12918-019-0683-1.

12. J. Nogales, B. Ø. Palsson, & I. Thiele, A genome-scale metabolic reconstruction of Pseudomonas putida KT2440: iJN746 as a cell factory. *BMC Systems Biology*, **2** (2008). https://doi.org/10.1186/1752-0509-2-79.

13. J. Nogales, J. Mueller, S. Gudmundsson, F. J. Canalejo, E. Duque, J. Monk, A. M. Feist, J. L. Ramos, W. Niu, & B. O. Palsson, High-quality genome-scale

metabolic modelling of *Pseudomonas putida* highlights its broad metabolic capabilities. *Environmental Microbiology,* **22** (2020) 255–269. https://doi.org/10.1111/1462-2920.14843.

14. R. Mahadevan, D. R. Bond, J. E. Butler, A. Esteve-Nuñez, M. V. Coppi, B. Ø. Palsson, C. H. Schilling, & D. R. Lovley, Characterization of metabolism in the Fe(III)-reducing organism *Geobacter sulfurreducens* by constraint-based modeling. *Applied and Environmental Microbiology,* **72** (2006) 1558–1568. https://doi.org/10.1128/AEM.72.2.1558-1568.2006.

15. J. Sun, B. Sayyar, J. E. Butler, P. Pharkya, T. R. Fahland, I. Famili, C. H. Schilling, D. R. Lovley, & R. Mahadevan, Genome-scale constraint-based modeling of *Geobacter metallireducens. BMC Systems Biology,* **3** (2009) 1–15. https://doi.org/10.1186/1752-0509-3-15.

16. A. M. Feist, H. Nagarajan, A. E. Rotaru, P. L. Tremblay, T. Zhang, K. P. Nevin, D. R. Lovley, & K. Zengler, Constraint-based modeling of carbon fixation and the energetics of electron transfer in *Geobacter metallireducens. PLoS Computational Biology,* **10** (2014). https://doi.org/10.1371/JOURNAL.PCBI.1003575.

17. J. M. Monk, C. J. Lloyd, E. Brunk, N. Mih, A. Sastry, Z. King, R. Takeuchi, W. Nomura, Z. Zhang, H. Mori, A. M. Feist, & B. O. Palsson, iML1515, a knowledgebase that computes Escherichia coli traits. *Nature Biotechnology,* **35** (2017). https://doi.org/10.1038/nbt.3956.

18. M. L. Mo, B. Ø. Palsson, & M. J. Herrgård, Connecting extracellular metabolomic measurements to intracellular flux states in yeast. *BMC Systems Biology,* **3** (2009). https://doi.org/10.1186/1752-0509-3-37.

19. P. Baloni, V. Sangar, J. T. Yurkovich, M. Robinson, S. Taylor, C. M. Karbowski, H. K. Hamadeh, Y. D. He, & N. D. Price, Genome-scale metabolic model of the rat liver predicts effects of diet restriction. *Scientific Reports,* **9** (2019) 9807. https://doi.org/10.1038/s41598-019-46245-1.

20. N. C. Duarte, S. A. Becker, N. Jamshidi, I. Thiele, M. L. Mo, T. D. Vo, R. Srivas, & B. Ø. Palsson, Global reconstruction of the human metabolic network based on genomic and bibliomic data. *Proceedings of the National Academy of Sciences of the United States of America,* **104** (2007) 1777–1782. https://doi.org/10.1073/PNAS.0610772104.

Index